RENEWABLE ENERGY

There is nothing new under the sun

Irrespective of our stance on climate change,
the future of renewable energy will be determined by
accessibility, affordability, and reliability

Solomon D Matios

Copyright @ 2024 by Rev Engineering

All rights reserved. No part of this book may be used or reproduced without the prior written permission of Rev engineering.

All streams flow into the sea, yet the sea is never full.
To the place the streams come from,
there they return again.

What has been will be again,
what has been done will be done again;
there is nothing new under the sun.

Is there anything of which one can say,
"Look! This is something new"?
It was here already, long ago;
it was here before our time.

No one remembers the former generations,
and even those yet to come
will not be remembered
by those who follow them

Ecclesiastes Chapter 4-11

The cultural and scientific accomplishments of the past
have allowed mankind to build an advanced and prosperous world.
This book is dedicated to the wise men and women who acknowledge the
contribution of past accomplishments to the achievements of today and going
forward.

Contents

New Energy Revolution 1
   Black gold 5
Environmental degradation 14
   Air pollution 14
   Land degradation 16
   Water pollution 19
   Extreme measures 21
   National Security 29
Understanding Renewable energy 33
   Accessibility 39
   Energy prices 40
   Employment 41
   Self-Sufficiency 45
Energy Efficiency 49
   Buildings efficiency: time-tested techniques 50
      Passive solar heating: orientation 50
      Passive solar cooling: material selection 54
   Buildings efficiency: New techniques 60
   Energy Star program 63
Hydro Energy 66
   Economics 72
   Variations 73
   The nuts and bolts of hydropower 84
   Hydropower benefits 89
Wind Energy 91

    Understanding wind energy 94
    Wind turbine types 104
    Wind farms 105
    Small-scale wind power 110
Active Solar Energy 113
    Types of solar panels 117
    Additional hardware 121
    Solar power installations 123
    Solar thermal (heat) energy 125
Geothermal Energy 130
    Source of geothermal energy 132
    Exploration 134
        Exploration Methods 137
        Extraction 139
    Geothermal Power 142
    Direct Applications 147
        Ground source heat pump (GSHPs) 148
        Agriculture 151
    Renewable and sustainable 154
Bioenergy 158
    Biopower for power utilities 158
    Biofuel for the transportation industry 159
    Bioproducts for the manufacturing industry 161
Conclusion 163
References 166
Photo credits 168

New Energy Revolution

> Is there anything of which one can say,
> "Look! This is something new"?
> It was here already, long ago;
> it was here before our time
>
> *Ecclesiastes Chapter 4-11*

The accomplishments of previous generations are ridiculed and forgotten only to be re-introduced as new in the future. Mankind is quick to acknowledge and praise new ingenuity, only to realize it was discovered long ago. It is said that history is studied to prevent mistakes from reoccurring. In this case, we will study history to benefit from the accomplishments of past generations that have come and gone long before our time. Most importantly, we will uncover past developments that have paved the way for our current technological advancements.

Today, we are amid the rejuvenation of ancient technology, mistakenly, considered new and technologically more advanced by many. This new technology has the potential to positively impact our modern way of life. It can revitalize the way we power our homes, business, and factories as well as the way we fuel our vehicles. It is the driving force behind our computing and machinery powers. In addition, this new technology promises to change our climate for the better while improving our public health in the process.

This mistakenly called "new technology" was familiar to ancient Egyptians who used it to sail their boats along the river Nile over 5000 years ago. Native American tribes of the past would recognize it because it was a concept they implemented in building their dwellings. Ancient Greeks, Romans, African, and Central Asian cultures made use of this technology to grind grain and irrigate their agriculture. Over 3000 years ago, the Chinese were familiar with the concept when they created concave reflectors to concentrate sunlight into more powerful rays for starting a fire. Almost all ancient cultures around the world used its powers for therapeutic & medicinal spas. Yet, this time-tested achievement continues to be labeled as new and modern in today's world. It is an old but recently re-discovered phenomenon. We know it as Renewable Energy. It is a combination of multiple sources of energy from infinite sources, such as the sun and the energy deep within the earth. Renewable energy promises to replace fossil fuel as the dominant source of energy gradually. In many media outlets, renewable energy is paraded as the beginning of a new revolution. A revolution driven by an urgent demand to curb an ever-increasing reliance on fossil fuels of coal, natural gas, and crude oil, and their impact on climate change.

The discovery of commercial-level fossil fuels and their integration into our economies have led to great achievements in history. Mankind has witnessed multiple industrial revolutions which have helped improve living conditions for many around the world. As a result, we have witnessed a booming population growth in the last few hundred years. According to the United Nations (UN), the global population has increased from nearly one billion to over seven billion in just 200 years. The skyrocketing population growth has been the driving force behind the record levels of energy consumption in recent decades. The transportation, agriculture, and pharmaceutical industries, to name a few, are all consuming energy at a record level.

Unfortunately, the growing consumption of fossil fuels is attributed to worsening climate change which is a major contributor towards deteriorating public health. Multiple world organizations, such as the United Nations (UN) and World Health Organization (WHO), as well as domestic entities such as the United States Center for Disease Control (CDC), have warned that deteriorating air quality, rising global temperature, and increased CO_2 emissions are on the rise. These changes in the environment contribute to public health deterioration, leading to critical health concerns such as heat stress, asthma, and malnutrition due to failures in agricultural output.

In addition, many claim the status quo will lead to a deteriorated environment that can lead to tragic health impacts, depleted natural resources, and possibly challenge humanity's existence on earth. For example, the exploitation of precious and reserved areas, such as national parks, ocean fronts, and arctic regions for fossil fuels, has led to the contamination of these irreplaceable territories. Furthermore, regional and global conflicts for access to these limited resources are on the rise. Therefore, fossil fuels are increasingly scrutinized and frowned upon despite our continued dependency on them. Unsurprisingly, climate change and the energy debate have dominated the airwaves for the past decades, often pinning public opinion against one another. The global movement to abort fossil fuels for cleaner, cheaper, and renewable energy sources like solar, wind, hydro, geothermal, and biofuel is front and center.

There are two sides to every story. If one side attributes fossil fuels to deteriorating climate and worsening public health, the other side credits fossil fuels for the gains in technological advancements and improved ways of life over the last 200 years. In fact, fossil fuels are still the main driver of global economic activities. For example, the cheapest source of electricity has been coal plants for decades. Currently, a different type of fossil fuel, natural gas, driven plants are one of the cheapest sources of electricity. Many proponents argue there is little evidence that fossil fuels lead to the above-mentioned challenges of climate change, deteriorating public health, global conflicts, and land contamination. They argue that there is not enough data for weather patterns or climate change analysis because data has only been available for the last 100 years or so. More time and more data are needed to study the topic to build a convincing argument against the status quo and change our energy consumption behavior.

Most importantly, many fossil fuel advocates point out the skepticism of the climate change agenda. Climate change research is guided and financed by a few wealthy nations. Thus, results can be manipulated, and analysis can be politically motivated. For example, the group of Western nations who are advocating for a reduction in fossil fuel consumption are the same group of nations who have benefited the most from it. It is the fossil fuel-based industrial revolution that helped these nations advance their economies and improve their populations' way of life. Is it possible that these nations have just come to learn the consequences of fossil fuels and want to correct past mistakes? Or is there a hidden agenda to stall the industrialization and economic growth of competitor nations? One reason for the skepticism is the lack of credibility these nations face.

The Western nations who are advocating to safeguard the environment for humanity are the same nations who have gone to great lengths to create the environmental degradation they claim is a threat to our existence. Their history of colonialism and human exploitation for profit tarnishes their reputation. Their unrelenting will to unleash military might, in the name of human rights & Democracy, Rights to protect (RTP), and Islamic terrorism, has not gone unnoticed. Military intervention around the world including in former Yugoslavia, Afghanistan, Iraq, Libya, Syria, and Ukraine, all rich in natural and strategic resources, have been disastrous for the locals, leading to the partitioning of nations such as Yugoslavia, dismantling of a nation-state, such as Libya, as well as refugee crisis like in Syria. Their past and present behavior makes these Western nations untrustworthy and unfit to preach to others.

Another reason for skepticism is the "do as I say, not as I do" policies implemented by these governments. For example, the Biden administration in the United States passed the Inflation Reduction Act (IRA) to stimulate the renewable energy industry with nearly $1 trillion, effective January 2023. Despite its purpose to promote renewable energy and fight climate change, the act will reward the oil and gas entities with access to nearly $500 billion in new federal funding and tax credits. This new revenue is intended to encourage oil companies to store carbon residues below ground, among many other projects.

The efforts will contribute towards a cleaner environment but reward these oil companies for continuing business as usual. For example, The IRA increases the tax credits for carbon capture from $35 per ton to $130 per ton. If these companies can keep the carbon stored permanently, they can be rewarded up to $150 per ton, compared to the $50 per ton credit that was available previously. The extra funding encourages the oil and gas companies to drill more fossil fuel rather than disrupt their operations. Predictably, Oil and gas CEOs are praising the IRA. BP America head David Lawler called the IRA "a key enabler for starting the transition" while ExxonMobil CEO Darren Woods praised it as a model for what climate policy should look like.

On the other side of the Atlantic, Germany was one of the most invested nations in renewable energy. Germany has made huge progress in increasing its electricity production from wind, solar, and overall renewable sources while decommissioning its coal plants. Unfortunately, the reduction in coal consumption that started in year 2015 has started to reverse its trend in year 2020. As the chart below illustrates Germany's coal consumption started trending upwards beginning year 2020. The shift in coal consumption is not a coincidence. In preparation for a possible natural gas supply disruption that may result from the Russia/Ukraine conflict, the Germans began restarting or extending the lifespans of many of their coal-based generators. It seems the German renewable energy policy was to pursue it when convenient. In other Western nations, such as the United States, a decline in coal has been replaced by increased use of another fossil fuel, natural gas.

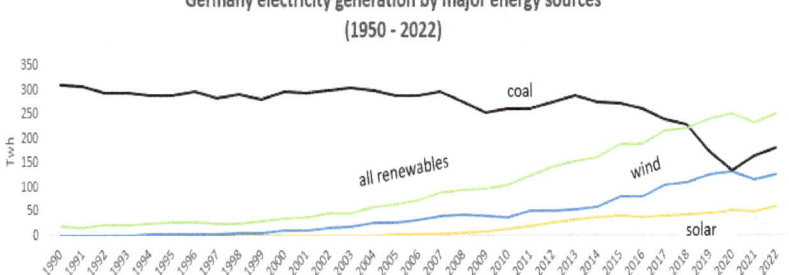

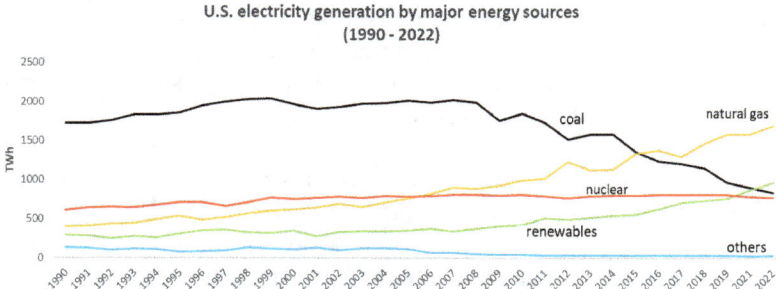

Globally, the use of coal has been and still is on the rise. The natural gas trend has been on the rise since the 1990s as well. The below chart illustrates the trends in the use of fossil fuels in the power utility sector over the past few decades. Despite the rhetoric and noise, our economies remain fossil fuels dependent.

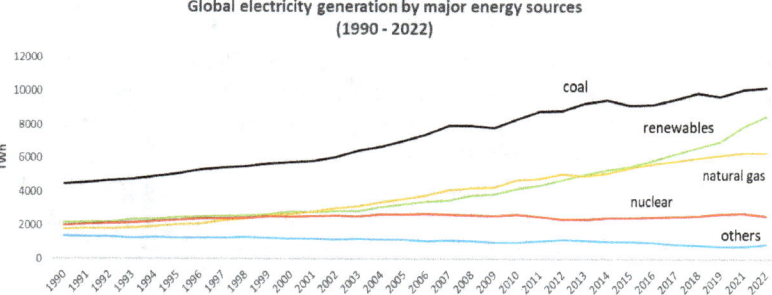

To gain an appreciation for both sides of the energy debate, a clear understanding of the historical background of both fossil fuels and renewable energy sources is a must. In addition, the impact of these resources on climate change and public health must be examined. Finally, the impact on our economies and way of life must be considered before advocating for a behavioral change in energy consumption.

Black gold

The term "fossil fuel" refers to the nonrenewable energy sources of crude oil, coal, and natural gas. Even though fossil fuels have only been used commercially since the late 1800s, they have a very long history going back thousands of years. Scientifically, fossil fuels are the result of vegetation; trees, plants, ferns, grasses, and animals remain buried deep underground for millions of years. These decomposed remains have a high carbon and hydrogen content. Under extreme pressure and heat, the carbon and hydrogen molecules break apart and turn into peat and plankton. With increased time and pressure from the earth/dirt above, the peat solidifies and turns into a coal seam while the plankton turns into natural gas and crude oil. Collectively, they are known as fossil fuels. Today, we mine coal and drill for oil and natural gas for their stored carbon and hydrogen molecules, which result in energy when burned.

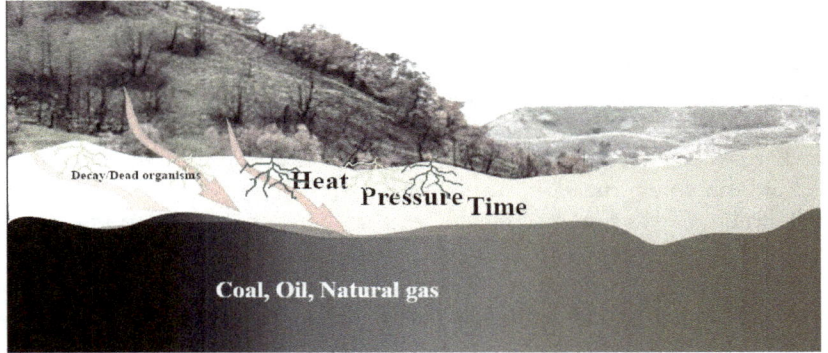

simplified process of the origin of fossil fuels

Although the extraction of commercial-level quantities is recent, a decent amount of fossil fuel has been used in different civilizations worldwide for thousands of years. Some references go back to biblical times when crude oil/tar was used for many purposes, including in construction materials. In Genesis 11:3, the bible describes how people in those days used bricks hardened with fire and tar (crude oil), found seeping on the surface of the ground, for mortar to build towers. The Bible also makes another reference to tar in Genesis 6:14 which tells the story of God instructing Noah to build a large boat from cypress wood and waterproof it with tar, inside and out.

Around 4000 BC, the Sumerians of Mesopotamia region in modern-day Iraq used asphalt to glue their mosaic works on floors and walls. Ancient Egyptians used to it mummify their dead. Persians, Native Americans, and Chinese all used crude oil for medicinal benefits as a skin treatment. In addition, crude oil was used for handles on arrows and knives and burn-weapons during warfare. A lot of these crude oil/tar-derived products have lasted into our modern era where they are produced at the industrial level. These modern products include plastics, waterproofing sealants, asphalts, and rooftops.

Another reference testifies to the times of Herodotus over 2000 years ago. Herodotus, an ancient Greek historian, also known as the father of European history, describes Ardericca, a military staging post, in the book "THE HISTORIES" in this way. "Ardericca is 210 stades from Susa, and it is also forty stades away from a well which is a source of three different products: bitumen, salt, and oil are all extracted from it." Herodotus was a well-traveled man and through his travel records, we get an in-depth understanding of the times when he lived from 485 BC to 425 BC.

A few centuries forward to the middle of the 19th century, people began to drill for crude oil purposefully. In 1859, Edwin L. Drake drilled the first commercially successful oil well in Titusville, Pennsylvania. The oil from these wells was refined to produce kerosene as fuel for fuel lamps. However, it wasn't until the invention of the first car with an internal combustion engine in 1885 by German engineer Carl Benz that crude oil was processed for other than kerosene. Carl Benz's car was powered by a byproduct of kerosene called gasoline. Not far behind, Henry Ford launched the Model T in 1908, an affordable mass-market automobile, in the U.S. The mass production of cars led to gasoline replacing kerosene as the most demanded refined product from crude oil. Additional oil discoveries in the Middle East contributed to oil becoming the most used energy source in the Western World in the early 20th century. In addition to oil demand for automobiles, both World War I & II created a demand for petroleum products as armies and navies expanded their fleets. Warships, tanks, and military trucks consumed a large amount of fuel and caused oil demand to soar worldwide.

In the late 19th century, we witnessed the rise of the Standard Oil Company, founded by John D. Rockefeller in 1870, as the dominant oil company in the United States. Standard Oil Company controlled nearly two-thirds of the petroleum industry in the United States. Following the antitrust lawsuits in 1911, Standard Oil was split into many regional companies. Standard Oil of New Jersey/Exxon, Standard Oil of New York/Mobil, and Standard Oil of California/Chevron would emerge as the dominant oil conglomerates in the years to come. Outside the United States, Royal Dutch Shell was established in 1907 following a merger between Shell Transport and Trading Company and Royal Dutch. Two years later, British Petroleum (BP) emerged from the Anglo-Persian Oil Company.

Today nearly 70% of the transportation industry is fueled by petroleum, generating trillions of dollars for those conglomerates. In addition to transportation, petroleum products are also used as feed for many consumer products, such as plastics, medical equipment, paint, and asphalt for roads and roof shingles.

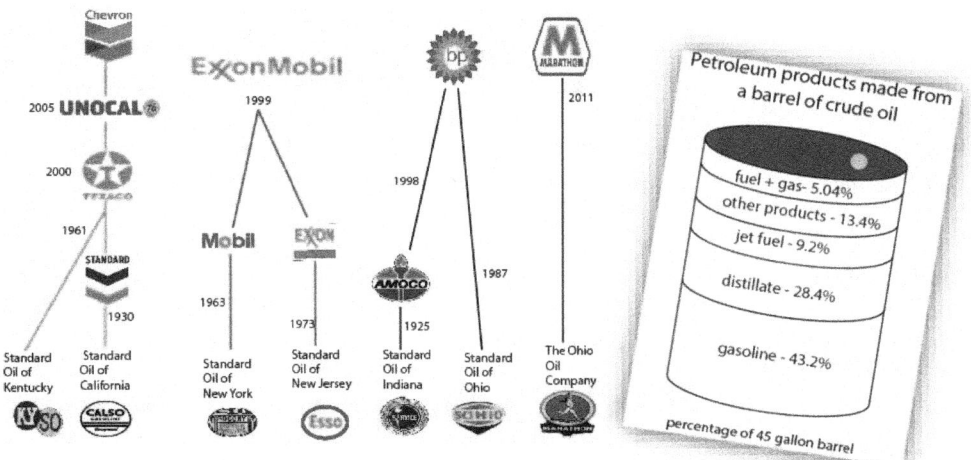

Petroleum was and still is the backbone of the transportation industry. Coal, on the other hand, was the backbone of the industrial revolution and later, a major source of electricity production. The Chinese were one of the earliest in recorded history to mine coal around 3000 BC. Marco Polo, on his trip to Asia in the late 1200s, described the use of coal in China;

> "It is a fact that all over the country of Cathay there is a kind of black stones existing in beds in the mountains, which they dig out and burn like firewood. If you supply the fire with them at night, and see that they are well kindled, you will find them still alight in the morning; and they make such capital fuel that no other is used throughout the country. It is true that they have plenty of wood also, but they do not burn it, because those stones burn better and cost less. Moreover, with the vast number of people, and the number of hot baths that they maintain--for everyone has such a bath at least three times a week, and in winter if possible every day, while every nobleman and man of wealth has a private bath for his own use--the wood would not suffice for the purpose."

Centuries later in 15th and 16th century Europe, coal became an increasingly important fuel for heating homes with firebrick chimneys.

The introduction of a more efficient steam engine in the mid-1770s made coal an essential energy source for industry and households. Coal-powered steam engines propelled the incredible powers of manufacturing and industrial activity and helped usher in the first industrial revolution. The industrial revolution that began in the mid-18th century elevated coal to the status of dominant energy source. Coal fueled the steam engine, which was the invention at the heart of the industrial revolution. Coal was available in abundance, and soon replaced water (which was the critical source of power before the Industrial Revolution) as the primary source of energy. Steam-powered machines made mass production possible. Coal also changed the way people traveled as steamships and steam-powered trains were burning coal to power boilers.

In the Americas, coal has an early beginning as well. Records indicate coal was an important fuel source for Native Americans in the 1300s. Coal was used for cooking, heating, and baking clay potteries. During the American Civil War of the 1860s, weapons factories started using coal in abundance. By the late 19th century, coke made from coal was the primary fuel for steel-making furnaces. During the same era, in the 1880s, utilities started using coal for electricity generation in the United States. Half a century later, by the early 1960s, coal had become the dominant energy source of U.S. electricity generation.

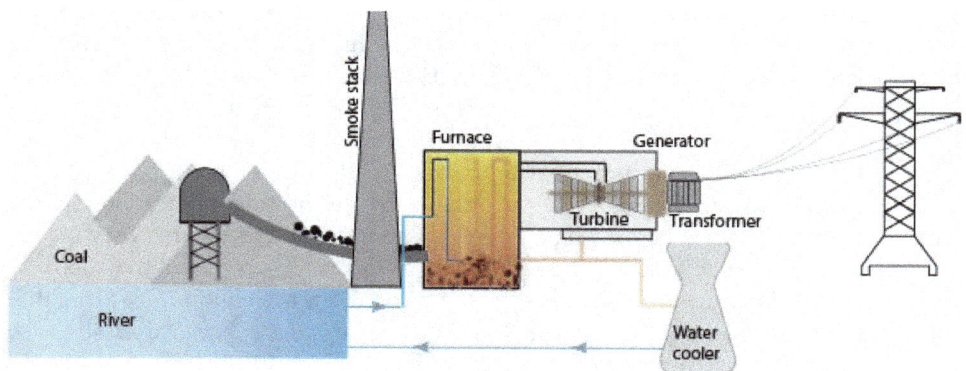

Currently, the United States has the largest reserve of coal in the world accounting for nearly 28% of global reserves. Most of that coal is located beneath the mountains of West Virginia (WV) in the Appalachian Mountains. The large quantity of coal in this area is a result of a natural process lasting millions of years. Experts testify WV was part of a vast coastal swamp covered by tropical forests around 300 million years ago. Over time, dead vegetation; plants and trees, sank to the bottom of the swamps and were covered by dirt and water. With increased time, pressure, and heat, they eventually turned to coal. According to a document in the Beckley Exhibition Coal Mine in Beckley, WV, every 12 inches of coal thickness represents 10,000 years' worth of plant material accumulation.

Additionally, the United States has large coal reserves in the plain fields of northeast Wyoming and southeast Montana known as the Powder River Basin (PRB). The PRB hosts some of the largest surface coal mines in the world. One of the open surface mines is the Eagle Butte Mine which covers about 400 acres of land and approximately 450 feet pit depth. Following the drilling, blasting, and removal of topsoil, nearly 50 feet of coal depth is accessed and mined. If 12 inches of coal thickness represents 10,000 years of plant material accumulation, then the 50 feet of coal depth mined at Eagle Butte Mine has taken over 40,000 years to make. Hence, it is clear why fossil fuels are considered non-renewable sources. Currently, there are 12 active coal mines in the PRB area contributing to the nearly 240 coal-fired power plants in the United States. As of 2022, coal is used in 10% of U.S. electricity generation, down from its peak of 60% a few decades ago.

An open coal mine in Wyoming, USA, hundreds of feet below the surface.

Throughout centuries, the use of coal grew steadily outpacing the use of wood for energy. After all, coal was in abundance, easier to transport, and gave off much more energy than wood. Between the late 1990s and early 2000s, coal-powered utilities generated nearly 60% of all electricity in the United States. In other parts of the world, coal has played a significant role in the development of economies. For example, coal accounted for 78% of China's electricity generation and nearly 60% of the country's total energy consumption a decade ago.

Today, the era of coal is in decline as it is targeted for extinction due to its pollutant nature. After all, coal is the most pollutant of all fossil fuels. For that reason, coal-powered utilities are in decline. In their place are natural gas-fueled plants, replacing one fossil fuel for another.

Recent advances in technology have allowed for natural gas to be as widely consumed as coal and oil. In previous decades, natural gas was burned or flared as a waste byproduct of oil drilling. Today, natural gas is the most used fossil fuel in electricity generation because it is much less pollutant. However, it remains a non-renewable, non-replenishable source.

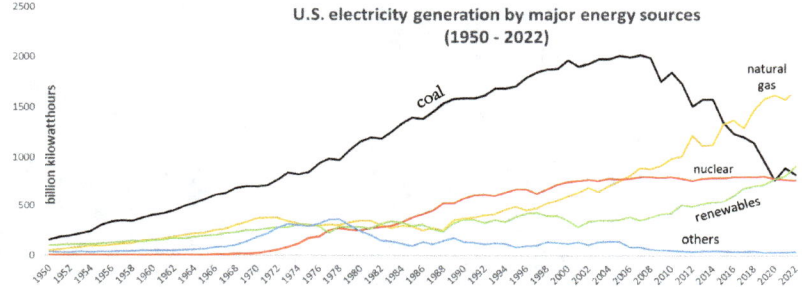

one of the few remaining coal-powered plants, Georgia USA

As previously stated, fossil fuels have been the dominant source of energy for over a century. During the 1700s, before the widespread use of fossil fuels, the world's population is estimated to be 670 million. By 2023 the world's population had reached over 7 billion. The exponential population growth is spearheaded by fossil fuel-driven industrialization and economic success. Many people around the world today enjoy the benefits of multiple industrial revolutions. For example, heavy-duty machines do most of the labor-intensive work, while intelligent devices help us perform better and quicker. Most of humanity benefits from improved health care and education which have resulted in child death rates decline and increased life expectancy in most parts of the world. Overall, mankind has achieved and as a result, benefited from the commercialization of fossil fuels.

Fossil fuels have contributed to all previous industrial revolutions of the past and today. Initially, fossil fuels were the backbone of the first industrial revolution, which had its origins in Great Britain in the 1760s. The first industrial revolution led to the evolution of economic advancement from agricultural to industrial. That era also introduced mechanical production and textile mills using coal-powered steam engines.

Coal/steam-powered train from the early 1900s, in operation in Asmara, Eritrea.

Following the First Industrial Revolution, the invention of the combustion engine, which was powered by gasoline, defined this era known as the Second Industrial Revolution in the late 1800s and early 1900s. In addition, the discovery of electricity along with developments in machines, tools, and computers, gave rise to the automation of factories which were all driven by gasoline or coal. The invention of planes and cars revitalized the transportation industry and increased fossil fuel consumption exponentially.

By the time the third industrial revolution started to emerge in the late 20th century, fossil fuels were firmly planted in nearly all human activities. Although this period is known as the age of the digital revolution, nuclear energy, and the electronics industry, it will most likely be remembered for the two world wars driven by heavy fossil fuel consumption. Military planes, tanks, and other equipment were fuel guzzlers.

The current, fourth industrial revolution, is considered the period of the Internet of Things, cloud technology, artificial intelligence, and virtual reality. This is also the era where we are witnessing a rapid shift from fossil fuel consumption towards renewable energy such as solar, wind, hydropower, biomass, and geothermal.

Despite the negative sentiment, fossil fuels still represent over 80% of global energy consumption. Chances are fossil fuels will remain the chief energy source for the near future. Many at the regional and local government levels warn about the loss of economic activity if fossil fuels-based economies are decommissioned too soon. Policies and regulations based on inconsistent and inaccurate research as well as misunderstood or misrepresented analysis will ultimately lead to harsh economic disasters. Fossil fuel-based factories and refineries will shut down, jobs will be lost, economic opportunities will be forfeited, and livelihoods will be destroyed. After all, fossil fuels have been the driving force behind every innovation made during the last two centuries, including in the manufacturing, transportation, agriculture, pharmaceutical, and other industries which all have contributed impressively to our modern way of life.

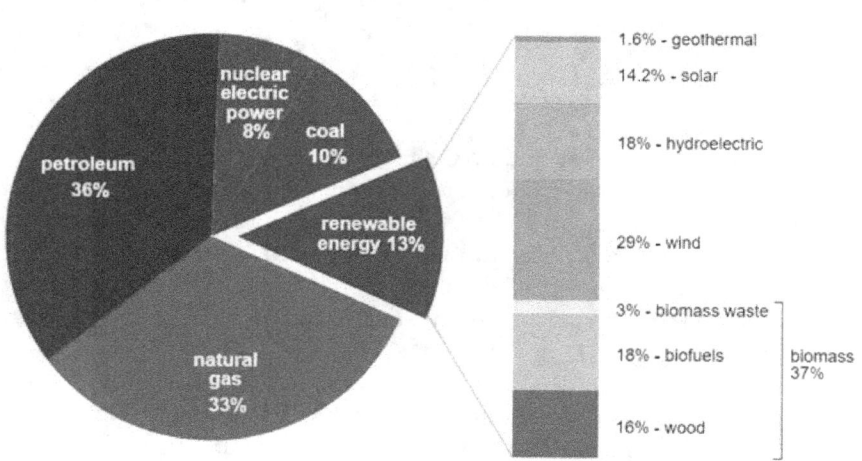

Environmental degradation

Despite our continued dependency, there is still an increasingly negative public sentiment towards the use of fossil fuels. Many scientists and institutions are convinced that the continued use of fossil fuels leads to irreversible consequences. In a recent COP conference, U.S. special presidential envoy for climate John Kerry addressed his counterparts in China, calling climate change a "universal threat" that will require all nations to come together despite any political differences. COP (Conference of the Parties) is an annual global summit hosted by the United Nations Framework Convention on Climate Change (UNFCCC). The conference is designed to bring together private and public businesses as well as world leaders to address climate change. More importantly, the goal is to accelerate climate change action by phasing out the use of fossil fuels.

As previously mentioned, John Kerry and the Western nations he represents have little credibility. Their energy policies do little to address climate change and reduce the use of fossil fuels. Regardless, billions of dollars in grants and loans are promised to developing economies to promote change in energy consumption behavior. There are even considerations for a carbon points system to credit nations who are aligned with the climate change agenda and punish those who are not. Are we headed toward the sanctioning of nations because of their energy policies similar to the military interventions of past decades in the name of "Rights To Protect" (RTP)? The following are some of the arguments made for the urgency to reduce or abandon fossil fuels and shift towards renewable, clean energy sources.

Air pollution

As previously explained, fossil fuels are the final products of dead organisms (animals & plants) buried deep inside the earth over millions of years. Under the right conditions of extreme heat, pressure, and an oxygen-free environment, these buried organic materials are turned into a solid rock where fossil fuels (coal, crude oil, and natural gas) are extracted from. These decomposed organisms contain high levels of carbon and hydrogen molecules, which are the source of fuel/energy. Burning these fuels creates a chemical reaction between the carbon, hydrogen molecules, and oxygen in the air in a process called combustion. Combustion breaks the bonds between those molecules, which results in the energy we consume. Unfortunately, there are side effects to the combustion process. Some of the carbon molecules mix with oxygen in the atmosphere to produce deadly byproducts of carbon dioxide (CO_2) and carbon monoxide (CO). These byproducts are the greenhouse gases that are causing much of the climate degradation. Since natural processes can only absorb about half of the amount of CO_2 that is emitted, there is a growth of net increase of atmospheric CO_2 annually. In the United States, the burning of fossil fuel, particularly from the power utilities and transportation industry, accounts for about three-quarters of carbon emissions.

In addition to carbon-based pollutants, burning fossil fuels releases other harmful agents. For example, coal-fired power plants singlehandedly generate about 40% of the dangerous mercury emissions and nearly 60% of the sulfur dioxide emissions in the U.S. Meanwhile, fossil fuel-powered vehicles are the main contributors of poisonous carbon monoxide and nitrogen oxide, which are the causes of smog. Some of the most harmful agents emitted when burning coal include:

> Sulfur dioxide (SO_2): Reacts with water and oxygen to form acid rain.
>
> Nitrogen oxides (NOx): Contributes to smog which leads to respiratory illnesses and cardiovascular effects. It is more dangerous for the elderly, young children, and people with asthma.
>
> Mercury and other heavy metals: When deposited in soil and water, they lead to neurological and developmental damage in humans and other animals.
>
> Fly ash and bottom ash: Residues created by coal-burning plants. They lead to lung cancer.

For more than a century, burning fossil fuels generated most of the energy required to fuel our vehicles, power our businesses, and keep the lights on in our homes. Even today, oil, coal, and natural gas supply nearly 80 percent of our energy demand. However, the consequences are believed to be detrimental to the environment and thus, to humanity. Pollution released into the atmosphere has a much broader global reach and impact on public health. For example, coal is a major part of the Chinese economy and fuels up to 78% of the nation's electricity generation and up to 60% of total energy demand. The consequences are, however, that many Chinese large cities are struggling with air pollution. Multiple studies have concluded that over 1 million Chinese die annually due to air pollution. Similar numbers are observed in other countries, such as India, which are heavily reliant on coal for their industries.

Coal-fired power plants are recognized from a distance by their smoke towers

Land degradation

In addition to air pollution, the process of extracting fossil fuels has a heavy toll on land and conservation. The required infrastructure to extract, process, and deliver fossil fuels to end-users leads to the destruction of conservation and surrounding communities. Mining, drilling, and unearthing large acreage of land negatively impact the landscapes and the wildlife that depend on them, as well.

For example, strip mining, or surface mining for coal, involves blasting away the entire landscape including mountaintops, to expose below-surface coal. Despite the best reclamation attempts at the end of the mine life, the land is never again suitable for previous native wildlife habitats. Such is the process involved at the largest U.S. surface mine in Wyoming's Powder River Basin, which was the source of nearly 60% of the coal mined in the United States in 2019.

all the extracting chemicals and waste products will accumulate to create an acid or toxic

Coal is also mined from underground using heavy machinery to cut coal from deep underground deposits. The process requires substantial financial investment and human health sacrifices to extract. It requires workers to be in an underground environment for an extended amount of time exposed to dangerous air particles. Over time, continued exposure to the coal dust causes scarring in the lungs, impairing the workers' ability to breathe. As a result, lung disease is widespread among coal workers.

In the case of crude oil, some is found in deep underground reservoirs where it is accessed by drilling using pumpjacks, which is a device used to extract crude oil from an oil well. Once at the surface, the crude oil is separated from any water and natural gas in the mix. The crude oil is then finally pumped into holding tanks before being transported to refineries. Pumpjacks can pump up to 10 gallons per stroke or up to 5 barrels per minute.

Pumpjacks are a safe way of extracting crude from underground. Their impact on the surrounding habitat is minimal. Unfortunately, the same cannot be said about other extraction methods. For example, some crude is located near the earth's surface in tar sands where it can only be accessed by strip mining. Tar sands are large deposits of bitumen or extremely heavy crude oil, consisting of a mixture of crude bitumen, silica sand, clay minerals, and water. Extracting this crude oil involves large-scale excavation of the land. The Athabasca deposit located in northeastern Alberta, Canada, is the largest known deposit of this kind in the world.

Athabasca deposit, located in northeastern Alberta, Canada

The oceans are not off-limit to resource exploitation as well. Thanks to modern technology, a large sum of crude oil is drilled from the oceans and seas using oil platforms. An oil platform, known as an offshore platform or offshore drilling rig, is a large structure with facilities for well drilling. The platform is used to explore, extract, store, and process petroleum and natural gas that lies in rock formations beneath the seabed in the continental shelf, as well as in lakes, inshore waters, and inland seas. Depending on the circumstances, the platform may be fixed to the ocean floor, consist of an artificial island, or float. Unfortunately, the level of contamination is unimaginable when these platforms fail. They can easily lead to shore contamination during oil spills, as witnessed in the BP oil disaster of the Gulf of Mexico in 2010 where over 130 million gallons of oil was spilled on the shores of the Gulf of Mexico.

Oil platform being built onshore on the coast of Texas, USA.

While some crude oil can easily be extracted using pumpjacks, surface mining, or offshore rigs, others are too difficult or expensive to extract. Thus, they require a different extracting technique, such as hydraulic fracturing, also known as fracking. Fracking is the process of extracting natural gas or oil from rock formations deep underground. It requires drilling down into the earth and injecting the newly drilled well with high-pressure fracking fluids which is a mixture of water, sand, and chemicals. The high-pressure fluid creates cracks in the deep-rock formation. When the hydraulic pressure is removed from the well, small grains of hydraulic fracturing proppants, such as sand, hold the fractures open and allows natural gas, petroleum, and brine to flow upwards more freely. A few days post drilling, the well is ready to produce oil or natural gas for years. The process can be carried out vertically or, more commonly, by drilling horizontally to the rock layer, to create new pathways or to extend existing channels. According to the U.S. Environmental Protection

Agency, fracking requires a large amount of water resources. For example, fracture treatments in coalbed methane wells use from 50,000 to 350,000 gallons of water per well, while deeper horizontal shale wells can use anywhere from 2 to 10 million gallons of water to fracture a single well.

Unfortunately, fracking is also highly controversial. The environmental and public health impacts outweigh its benefits. Groundwater and surface water contamination are significant concerns. Residents who live near fracking projects have suffered health issues, such as pregnancy and congenital disabilities, migraine headaches, chronic rhinosinusitis, severe fatigue, and asthma exacerbations. Fracking is also believed to initiate the triggering of earthquakes. For these reasons, fracking is restricted in some countries while completely banned in others.

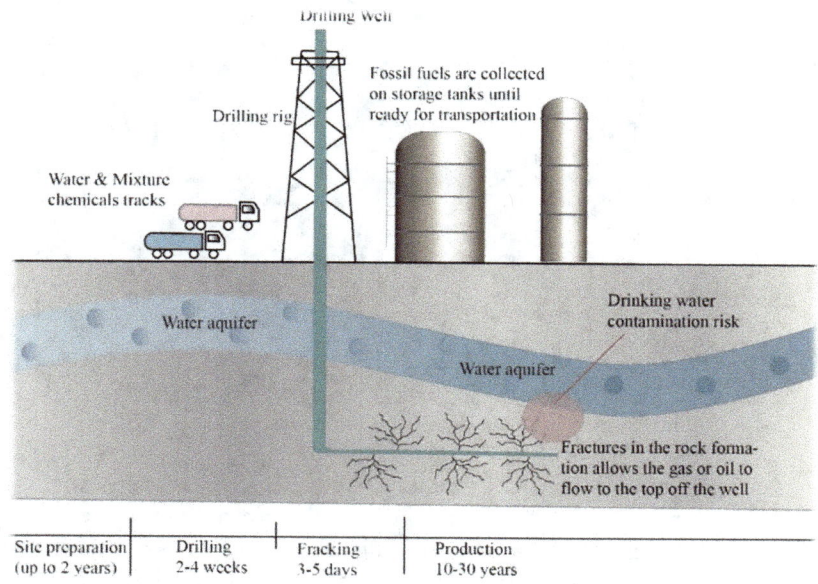

Simplified fracking process
It requires over 1400 truck trips to transport 2 to 5 million gallons of water.

Water pollution

In addition to air pollution and land degradation, the extraction of fossil fuels poses an undeniable threat to surface water and groundwater. The process generates wastewater, which can be laden with heavy metals, radioactive materials, and other pollutants. For example, acidic chemicals runoff into streams, rivers, and lakes during the mining process. Furthermore, transporting oil is susceptible to spills leading to the contamination of drinking water sources and jeopardizing entire freshwater or ocean ecosystems. Since the start of the Industrial Revolution, the ocean has become 30 percent more acidic. The increased acidity of the oceans impacts entire food chains and the coastal communities that rely on them. The oceans compromise two-thirds of the earth's surface and absorb as much as a quarter of all carbon that is emitted to the atmosphere. Changing their chemical characteristics has enormous consequences that can be felt globally.

Sometimes, the contamination of air, land, and water occur at the same time, in the same area

The process of extracting fossil fuel is an extremely dangerous process for both the people involved and the natural habitat of the area. Industrial explosives used to blast-open mining sites completely change the landscape beyond recovery. Chemicals used in the mining and drilling processes produce acid byproducts contaminating the land and air. When these acidic chemicals mix with rainwater or underground water, they lead to acid lakes. All these contaminations lead to health problems for workers and people living near mines. Below are some of the toxic chemicals used in mining and drilling fossil fuel and their impact on public health:

Sulfuric acid: a byproduct of many kinds of mining operations. It mixes with water and heavy metals to form acid mine drainage or acid lakes. In addition, the sulfur oxides emitted from burning coal react with moisture in the air to produce acid rain. Acid rain or any contact with sulfuric acid causes skin burns, blindness, and death.

Hydrochloric acid (HCl): used in fracking as part of a mixture of water, proppants, and chemicals. The mixture is pumped into the rock or coal formation to dissolve some of the rock materials, clean out pores, and enable gas and fluid to flow more readily into the well. Like most acids, HCl is corrosive and causes damage to the skin upon contact. In addition, inhaling HCl causes eye, nose, and respiratory tract irritation and inflammation, to name a few. There have been many incidents of groundwater contamination with HCl and other chemicals.

Ammonium nitrate and fuel oil (ANFO): ANFO is a mixture of 94% ammonium nitrate (AN) and 6% fuel oil (FO). It is a highly explosive compound widely used in blasting open-pit coal mining.

Extreme measures

Many Western nations have relied on military dominance to destabilize and exploit resources from many parts of the world, including the Middle East, Africa, and South America. As these resource-rich regions begin to nationalize their resources for the benefit of their people, many are forced to exercise extreme measures. One of those extreme measures is mining and drilling in areas that have been off-limits for resource exploitation. These areas include nature preserves and national parks, such as Theodore Roosevelt National Park in the northern part of the United States.

Theodore Roosevelt National Park, named after one of the country's great past leaders, Theodore Roosevelt, is a magnificent work of nature. The park is in the U.S. State of North Dakota. It encompasses over 100 square miles of scenic drives, foot, and horse trails, as well as hiking and camping. According to U.S. National Park Services, it is also home to much wildlife, including bison, bighorn sheep, white-tailed deer, mustang horses, elk, and over 186 species of birds.

A few years after the park was established in 1947, oil was discovered in the Bakken shale formation near the national park, bordering Montana, North Dakota, and the Canadian territories of Saskatchewan and Manitoba. The Bakken oil field is the largest continuous oil field in the world. Estimates from a 2013 United States Geological Service (USGS) survey showed that the Bakken oil field could produce up to 11.4 billion barrels of recoverable oil. However, it was economically unprofitable to extract using previously known methods. The introduction of hydraulic fracturing/fracking technology changed the fate of this majestic territory. By the end of 2014, the Bakken oil field was producing nearly one million barrels per day, contributing to 10% of U.S. total oil production, second only to Texas.

Despite the benefits of additional oil sources to the nation and the economic benefits of new jobs and tax revenues to the local economy, the disastrous impact on the park is unmistakable. The development of the area occurred so quickly that the long-term social and environmental costs were unforeseen. Although the park is protected from oil drilling, the land just outside its boundaries is not. Thus, the park was drastically impacted by the infrastructure and drilling activities that took place in its surroundings.

According to the National Parks Conservation Association (NPCA), rapidly increased production in the region led to air pollution, water pollution, oil spills, illegally dumped fracking waste, habitat fragmentation, heavy traffic, impaired views, and a host of social impacts. Fracking also produces large volumes of natural gas from the same oil wells. With no infrastructure to process or carry away that natural gas, oil companies had chosen to either leave it mixed in with the oil and load it onto trains or burn/flare the potent greenhouse gas into the atmosphere. For a very long time, North Dakota was flaring about a quarter of the gas produced via fracking. As many describe it, a state once known for its sparse population and ranching is now a giant oil industry playground. Fracking has allowed oil companies to drill for oil in areas that were unthinkable with traditional vertical drilling.

Extracting oil from isolated and nationally protected territories is half the battle. Transporting the oil to the population centers is another challenge. Motivated by high oil prices, the oil industry was committed to getting the oil out of the Bakken ground and to customers as fast as possible. Since there are no oil refineries or transport hubs near the Bakken oil fields, transporting oil by rail became the only option, leading to what is known as the "Bakken oil-by-rail" boom. Transporting oil by rail resulted in many oil spill accidents. For example, on July 6, 2013, one of the unit cars on a train carrying 77 tank cars, full of highly volatile Bakken oil, derailed and exploded in LacMégantic, Quebec. The train was on its way from North Dakota to the Irving Oil Refinery in New Brunswick. The incident destroyed nearly half of the downtown center and spilled a large amount of oil as it burned for days. Most importantly, 47 people lost their lives.

To combat the challenges of oil transportation, the U.S. Army Corps Engineers (USACE) engineered an underground pipeline to transport the Bakken oil to additional refineries near Patoka, Illinois. Known as the **Dakota Access Pipeline**, the pipeline was designed to carry up to 570,000 barrels per day of crude oil beneath the Missouri and Mississippi Rivers on a journey of 1,172 miles. This project was alarming, especially considering a report from the Pipeline and Hazardous Materials Safety Administration (PHMSA). The report cited more than 3,300 incidents of leaks and ruptures of oil and gas pipelines since 2010. The pipeline route is even more of a concern for native American tribes in that region. For example, the Missouri River is the primary drinking water source for the Standing Rock Sioux tribe in the central part of North and South Dakota. Additionally, the construction path threatens many ancient burial grounds and cultural sites held sacred by the Sioux Nation and other neighboring Native American tribes.

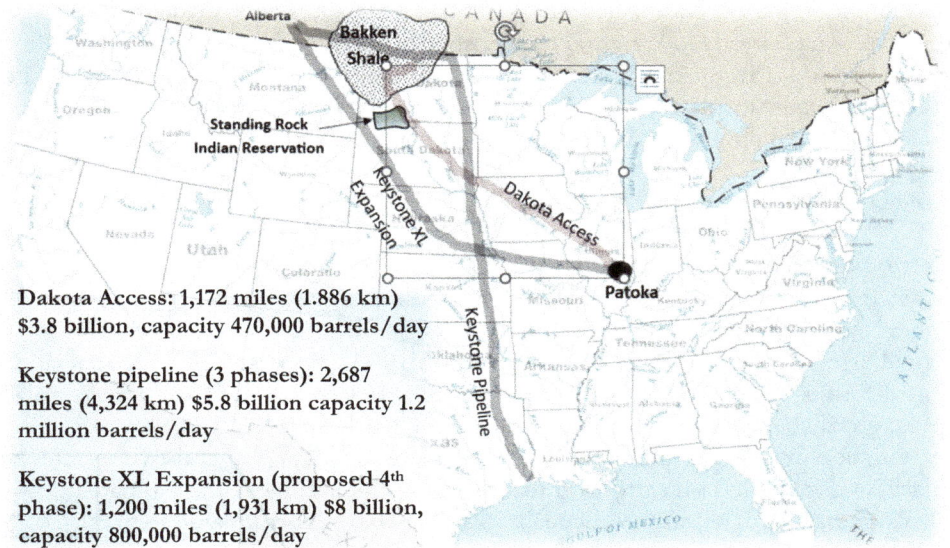

Dakota Access: 1,172 miles (1.886 km) $3.8 billion, capacity 470,000 barrels/day

Keystone pipeline (3 phases): 2,687 miles (4,324 km) $5.8 billion capacity 1.2 million barrels/day

Keystone XL Expansion (proposed 4th phase): 1,200 miles (1,931 km) $8 billion, capacity 800,000 barrels/day

On January 24, 2017, President Donald Trump signed an executive order reversing the Obama legislation and advanced the construction of the pipeline. The executive order expedited the environmental review, which Trump described as an "incredibly cumbersome, long, and horrible permitting process." On February 7, 2017, the Trump administration authorized the Army Corps of Engineers to proceed, ending the environmental impact assessment and the associated public comment period. The pipeline was completed by April and its first oil was delivered on May 14, 2017. Three years later, in July 2020, a District Court judge issued a ruling for the pipeline to be shut down and emptied of oil pending a new environmental review, only to be overturned by a U.S. appeals court a month later. At the time of writing this book, the Biden administration has managed to shut down the pipeline pending further investigation. If there is a lesson to be learned from this chaotic process, our continuous dependency on fossil fuels has become environmentally, socially, and politically toxic.

Following the coronavirus epidemic and oil price drop at the beginning of 2020, oil production from the Bakken oil fields of North Dakota fell by almost 50% to 827,000 barrels per day. Operating oil wells have dropped from their March 2020 peak of over 13,000 to below 800 in a short amount of time. The estimate for the break-even oil price for drilling Bakken wells ranged from $38 to $60 per barrel. At the recent fluctuating oil prices, North Dakota may have already seen its best oil production years. Oil companies and Wall Street financiers have made their quick profits and are already looking to other unspoiled areas, such as the Arctic wildlife refuge in Alaska. In the meantime, the re-conservation process must start. The large volumes of radioactive waste produced from fracking and the abandoned oil wells must be cleaned up. Unfortunately, North Dakota is using its own budget to deal with the aftermath. Recently, the state announced to use $66 million in federal funds designated for coronavirus relief, towards the cleanup efforts.

Arctic National Wildlife Refuge (ANWR) is another precious national park targeted for oil exploitation. ANWR is the largest national wildlife refuge in the United States encompassing over 19 million acres of northeastern Alaska. The refuge first became a federally protected area in 1960. This vast area of coastal lands, boreal forests, and alpine tundra supports many species of plants and animals. Polar bears, grizzly bears, black bears, moose, caribou, wolves, eagles, lynx, wolverine, marten, beaver, and migratory birds rely on this refuge. It is one of the finest, still intact, landscapes left on earth. A threat to this wildlife refuge is the estimated 7 to 11 billion barrels of oil stashed beneath it. A mere amount compared to Saudi Arabia's oil reserves of over 87 billion barrels plus many more untapped proven reserves.

In December 2017, Congress passed the Trump administration's Tax Cuts and Jobs bill, which included a backdoor provision to approve lease sales for drilling in the refuge. This move allowed for oil and gas rights to be auctioned off in the heart of one of the world's most iconic wild places. Since the beginning of the Trump administration, over a million acres have been leased. According to the U.S. Bureau of Land Management, the millions of acres the Trump administration has offered for oil and gas drilling included sensitive wildlife habitat around the Teshekpuk Lake. The area is one of the largest and most ecologically significant wetlands in the world.

Additionally, the administration put efforts to open an entire 1.5 million acres of coastal plain for gas and oil exploration in September 2019. This area was added to the refugee-protected status in 1980 by Congress because it is home to hundreds of species of birds, polar bears, and the Porcupine caribou herd, which is a vital resource for the native Gwich'in people. Fortunately, regulators have yet to finalize the environmental review process and the scheduled lease sale for 2019 has been postponed.

Unsurprisingly, the harsh weather conditions of the Arctic make drilling a risky business. The U.S. Department of the Interior has concluded a 75 percent chance of a major oil spill if oil production resumes in the area. Unfortunately, the studies have not stopped major companies like Shell and Exxon from aggressively pursuing a new "oil rush" in the Arctic Ocean.

The Trump administration initially projected that leasing would generate $1.8 billion in revenue over a decade, but it has subsequently cut that estimate in half. Some local politicians hailed the decision to lease for oil exploration as an economic boost for their state. They boasted that the oil boom would lead to new jobs and support economic growth and prosperity. However, the decision is short-sighted for short-term gain considering the plummeting oil prices and the availability of other options, such as renewable energy sources. Major U.S. financiers like Goldman Sachs and JPMorgan Chase have already publicly announced their discontent and promised not to fund oil and gas projects in the Arctic refuge.

Regarding offshore drilling, President Obama issued an executive order to ban drilling in the Arctic offshore waters permanently. As expected, the Trump administration issued an executive order overturning the decision. Trump's decision was later overruled by Alaska federal Judge Sharon Gleason for being unlawful. A small victory for many in a long, messy war. There is no convincing argument for drilling in one of the most precious habitats on earth at a time when the demand for fossil fuel is on a decline. The United States, which is the largest economy in the world, should be able to generate $1.8 billion in revenue over a decade by other means without tampering with one of the most precious and protected wildernesses on earth.

On the other side of the globe, the Russians and Europeans are a threat to the Arctic as they also strive to exploit the region for oil. In fact, oil exploration has already begun in parts of the Russian Arctic. Russian oil giant Gazprom had begun producing oil from the region as early as 2013. The Arctic is one of the world's fastest-warming regions, according to many scientists. This surge in climate change is destabilizing permafrost which has remained frozen year after year. As the permafrost destabilizes, so will the infrastructure (buildings, oil and gas pipelines, roads, railway, and military bases) built on top of it. The incident experienced by metals giant Norilsk Nickel, the world's leading nickel and palladium producer, in May 2020 is a reminder of the consequences of this risky endeavor. Melting permafrost shifted the foundation and ruptured a fuel reservoir at its power station about 200 miles north of the Arctic Circle. The incident resulted in the spilling of 21,000 tons of diesel fuel into a fragile ecosystem of rivers and wetlands.

According to a senior Russian official at the time, the oil started leaking on May 29th, 2020. Over 21,000 tons of oil had been released, contaminating the Ambarnaya River and surrounding subsoil. The oil spill had already polluted a large freshwater lake. The spill had also spread 12 miles (20km) north of the collapsed fuel tank nearing the Arctic Ocean. It is believed that the storage tank near Norilsk sank because of melting permafrost, which weakened its supports. The Arctic has had weeks of unusually warm weather. The oil spill is considered the worst accident of its kind in modern times in Russia's Arctic region.

Norilsk, Russian Arctic Circle

National parks and wildlife refuges are of significant importance. At the least, they are reminders of our planet's beauty to be cherished and admired by future generations. Unfortunately, our appetite for oil is exposing these areas to further resource exploitation and irreversible environmental degradation. There may have been a necessity or desperation to explore these precious territories for oil in previous decades. That desperation should not exist today. There are cleaner, cheaper, renewable, and more efficient options. Some may argue that resources/oil can be extracted from anywhere in the world efficiently and responsibly. However, regardless of our best efforts, equipment can get old and damaged, people can make

Renewable Energy Environmental Degradation

mistakes, and sometimes the weather takes a turn for the worst. Below is a list of oil spill disasters across the world. These incidents can serve as reminders of the consequences we face when we start to explore for oil in unchartered territories.

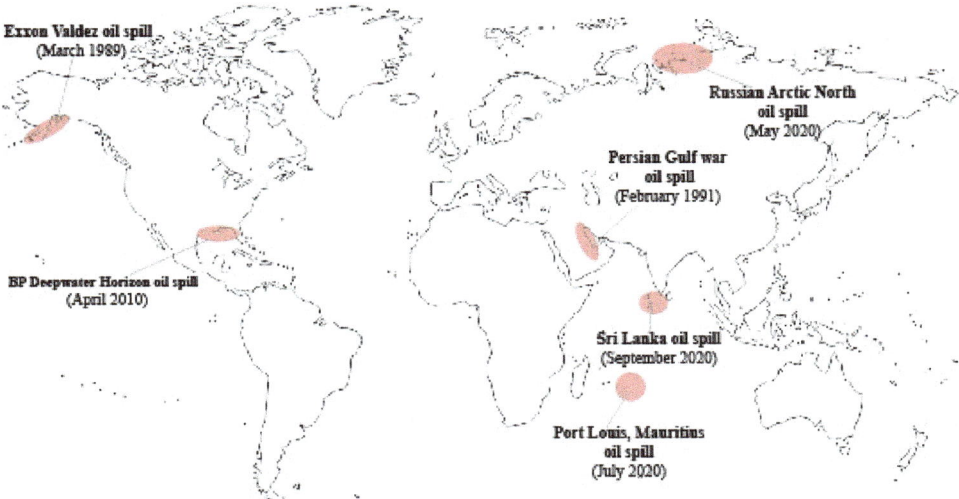

Exxon Valdez oil spill: On March 24, 1989, the oil tanker Exxon Valdez, bound for Long Beach, California, struck a reef and spilled nearly 11 million US gallons of crude oil in Prince William Sound, Alaska. The Valdez spill is the second largest in U.S. waters and is considered the most environmentally damaging oil spill worldwide. The oil affected 1,300 miles (2,092 km) of coastline. Exxon spent US$2 billion on cleanup efforts but only recovered less than 7 percent of the oil spilled.

BP Deepwater Horizon oil spill: The largest accidental oil spill in history occurred in the Gulf of Mexico on April 20, 2010. A surge of natural gas blasted through a cement well cap traveling up the rig's riser to the platform. The accident killed 11 workers and injured 17 more. Environmentally, over 130 million gallons of oil were released (according to the findings of the U.S. District Court) and about 1,300 miles (2,092 km) of the U.S. Gulf Coast from Texas to Florida were coated with oil. The oil platform finally capsized several months later, on September 17. In the lawsuits that followed, BP was found to be the responsible party and ordered to pay $65 billion in compensation to people who relied on the gulf for their livelihoods.

Port Louis, Mauritius: The bulk carrier MV Wakashio ran aground on a coral reef off the southeastern coast of Mauritius on July 25, 2020, spilling over 1,000 tons of oil and threatening a protected marine park boasting mangrove forests and endangered species. Three weeks later, on August 15, a crack in a cargo-holds at the vessel's stern forced the ship to break into two. Mauritius declared an environmental emergency and salvage crews raced against the clock to pump the remaining 3,000 tons of oil off the stricken vessel.

Sri Lanka: Nearly a month after the Port Louis oil spill, the New Diamond supertanker was burning for over a week, 34 miles off the coast of Sri Lanka. The supertanker, carrying the equivalent of about 2 million barrels of oil, had been transporting its cargo from Kuwait to a port in India when the fire broke out on Sept 3, 2020. Most of the fuel oil from the MT New Diamond was quickly contained, averting a major environmental and economic disaster for the 2 million Sri Lankans who depend directly on coastal fisheries and the Sri Lankan tourism industry. More than 80% of hotels in Sri Lanka are built along the coast.

National Security

In addition to environmental and health concerns, the dependency on fossil fuels is also a cause of growing numbers of global conflicts. Rising population and increased consumerism are leading to unsustainable demand for fossil fuels. Since fossil fuels are finite resources, their limited availability has been a source of conflict for many years. According to the United Nations Environment Programme (UNEP) 2010 report, over the past 60 years, 40 percent of civil wars have been associated with natural resources including oil. Additionally, there have been at least 18 violent conflicts fueled or financed by natural resources since 1990.

A good example of a resource-rich nation ravished by internal and external conflicts is the Democratic Republic of Congo (DRC) in southwest Africa. According to the Secretary-General's Special Representative for the Democratic Republic of Congo (DRC), Martin Kobler, acknowledgment in 2014, the exploitation of natural resources had fueled the extensive conflict that has ravaged the country and taken millions of lives. The extraction of minerals such as Colton and Cassiterite for the electronic industry as well as gold, timber, and oil has fueled both internal and external conflicts.

The DRC is the 2nd largest country in Africa and the 11th largest in the world. It comprises an area as large as all Western Europe. The DRC is also arguably the richest country in the world when it comes to natural resources. It is blessed with all kinds of resources such as copper, gold, diamonds, cobalt, uranium, coltan, rubber, oil, and thick forests. In addition, limitless water, from the world's second-largest river (The Congo), great climate, and rich soil make the DRC agriculturally rich as well. Unfortunately, the DRC is also home to the world's bloodiest conflict since World War II. Over five million people have died and millions more have been driven to the brink of starvation and disease. Since gaining independence in 1960, there have been countless armed conflicts involving over a hundred internal and external armed groups in the country.

All the country's commercial ports and political power seats are on the western side of the country along the Congo River and the Atlantic Ocean. Yet, all its armed conflicts and military unrest have been, and still are, in the less populated eastern region. It is also no secret that most of the country's resources are in the eastern region of the country. Unfortunately, wherever there are resources, armed conflicts are almost guaranteed. After all, all wars are resource wars.

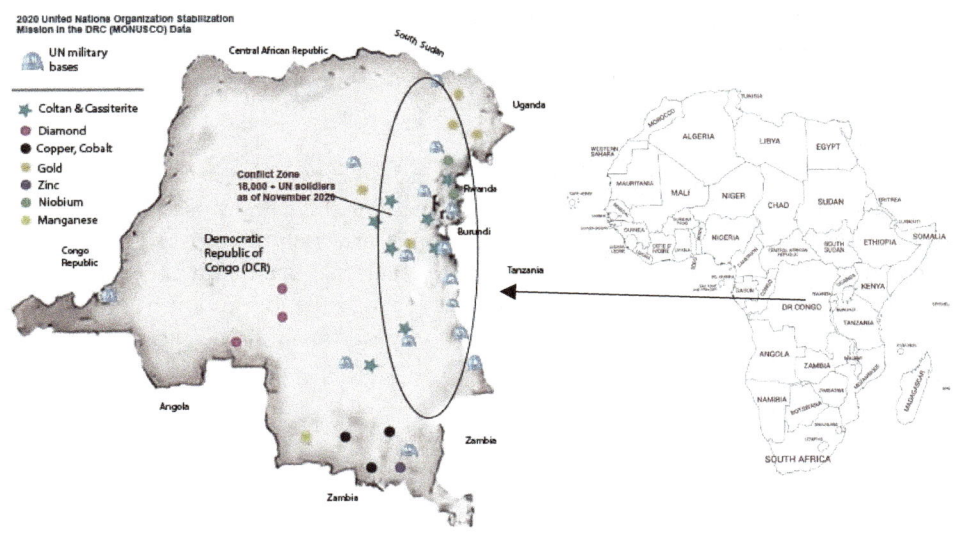

Ever since the Europeans' arrival on its shores, the DRC has been exposed to slavery, death, brutality, and impoverishment. When Portuguese traders arrived from Europe in the 1480s, they realized they had stumbled upon a land of vast natural wealth, including natural and human resources. Soon, the Congo was home to a supply of slaves. Over four million people were forcibly shipped to the Americas on English ships. Late in the 19th Century, the interior of the Congo was opened by the British-born explorer Henry Morton Stanley. Soon after, King of the Belgians, Leopold II, hacked a vast part of the country as his private empire. Under his control, millions of Congolese perished. Today the DRC remains one of the poorest in the world despite the abundance of natural resources. Its citizens have not experienced peace and stability for centuries. Unfortunately, the DRC's resource blessings have turned into a resource curse.

Currently, since oil is the most sought-after commodity, any region in the world with an abundance of oil reserves will certainly experience the ravages of war similar to what the DRC experienced and continues to experience. As the most prosperous oil region in the world, the Middle East is and has been a hotbed of resource conflicts for over a century. Commercial oil was first discovered in the region in the early 1900s. Anglo-Persian Oil Company (APOC), later British Petroleum (BP), discovered and then started producing oil in Iran by 1911. Following World War I, APOC found more oil in neighboring Iraq. In 1932 Standard Oil Company of California discovered oil in commercial quantities in Bahrain, followed by discoveries in Saudi Arabia in 1938. World War II delayed the development of whatever fields had been discovered in the 1930s until the 1950s. During the 1950s and 1960s, additional regional countries, including Kuwait, Qatar, and Abu Dhabi, started producing and exporting oil.

A few decades later in 1951, BP's investment in Iran came under attack following the election of a new prime minister, Mohammad Mosaddegh. Mosaddegh introduced a range of social and political reforms, of which the most significant was the nationalization of the Iranian oil industry, which had been built and operated by the British since 1911 through APOC/BP. Bitter over the loss of control of the Iranian oil industry and determined to keep cheap oil flowing to Western economies, the British along with the help of the U.S. government, overthrew Prime Minister Mosaddegh in a bloody coup in 1953. Iran's new government soon reached an agreement with foreign oil companies to restore the flow of Iranian oil to world markets in substantial quantities, giving the United States and Great Britain the lion's share of the restored British holdings. In return, the U.S. massively funded the new government of Mohammad Reza Shah until 1979, when the regime was toppled by a revolution that created an Islamic Republic. The revolutionary government or Islamic Republic then nationalized Iran's oil reserves for the second time. Relations between Iran and Western countries have been toxic since.

Thousands of miles to the west, the Arab-Israel conflict tested the relationship between Western powers and the Middle Eastern nations. In retaliation for Israel's invasion of Arab lands and retaliation for U.S. assistance to Israel's war efforts, the Organization of Petroleum Exporting Companies (OPEC), which was made up of primarily Middle Eastern nations, agreed to cut back their oil production and cut off oil supply to the world in 1970. Widely known as the "1970 oil embargo", the weaponization of oil by OPEC members brought many world economies to their knees. The oil embargo of 1970 and the 1979 Iranian revolution led to oil market fluctuations that crippled the economies of the U.S. and most Western nations.

In the decades to follow, the U.S. relied on its military dominance to secure resources worldwide. For example, in 2001, the United States' ambition to topple seven regimes, including Iran, Iraq, Syria, Lebanon, Libya, Sudan, and Somalia, was confirmed by former U.S. General Wesley Clark years later. All seven countries have a considerable amount of oil wealth or serve as strategic oil shipping lanes. In recent years, the U.S. has maintained direct or indirect military presence from the mountains of Afghanistan to the deserts of Saudi Arabia, Iraq, and Iran and the shores of Syria. The region that recorded some of the earliest resource conflicts over freshwater 6000 years ago continues to suffer from another resource conflict over oil.

Similarly, the South American nation of Venezuela continues to struggle with internal and external pressures for its vast oil resources. After all, Venezuela has the largest oil reserves in the world. It is believed numerous coups and assassinations

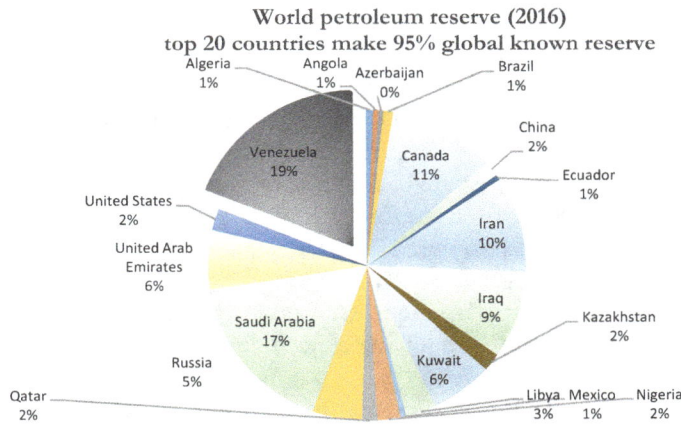

have been attempted on its past and present leadership. In addition, there have been multiple attempts to seize the nation's assets. According to Reuters, Jan 19, 2024 report, A U.S. court approved claims by 17 Venezuela-linked creditors, including ConocoPhillips, Rusoro Mining, and Koch Industries, to get proceeds from a coming auction of shares in the parent of Venezuela-owned oil refiner Citgo Petroleum. The auction will be held to satisfy claims worth $20.8 billion for expropriations and debt defaults. Citgo Petroleum is a Houston-based refiner and operates an 807,000-barrel-per-day refining network, 38 terminals, six pipelines, and supplies 4,200 independent retailers. The auction, which could lead to one of the largest court sales in U.S. history, was launched by U.S. Judge Leonard Stark after he received a green light from the U.S. Treasury. Understandably, Venezuela's foreign affairs ministry described the auction as a new aggression against the South American nation's assets.

Many other countries, such as Sudan and Ecuador, as well as territories like the "South China Sea" remain hotbeds of conflicts for oil and natural gas. Even one of the most isolated areas of the globe, the Arctic, has become a conflict zone between the United States, Russia, and European nations. Armed conflicts will certainly increase as the demand for fossil fuels increases. Undoubtedly, a shift from a fossil fuel economy towards locally available, infinite, and renewable sources of energy can help minimize global conflicts. Investing in Solar, wind, hydro, and geothermal energy can play a vital role in energy self-sufficiency for many regions. After all, every nation and every region has access to abundant energy from the sun, the wind, and from beneath the earth's surface.

Understanding Renewable energy

The fossil fuel industry played a significant role in economic success worldwide. As previously mentioned, fossil fuels are the backbone of multiple industrial revolutions that have elevated the development of modern economies. Investments in the fossil fuel industry have created economic opportunities and millions of jobs for many for the past 200 years. Therefore, a shift from fossil fuels to other sources of energy will be undoubtedly an uphill task. However, if the renewable energy industry can address the concerns of climate change, economic opportunities, global conflicts, and self-sufficiency, adapting to renewable energy sources may prove more beneficial. In addition, the following expectations must be fulfilled.

> Renewable energy must be **abundant and easily accessible** without the plundering of the little remaining conservation or nature preserves.
>
> Renewable energy must be **affordable** and allow many to participate and benefit from its economic fruits.
>
> Renewable energy must help nations achieve **self-reliance** which will help minimize global resource conflicts or resource exploitation.

The term renewable energy is interchangeably used with alternative energy or clean energy. For discussions in this book, "alternative energy" or "clean energy" refers to non-fossil sources of energy that do not emit greenhouse gases into the atmosphere. These energy forms include sources from nuclear, solar, wind, biofuel, hydro, and geothermal, to name a few. On the other hand, renewable energy refers to any clean energy source that is infinite or can be replenished in a short amount of time. Thus, renewable energy is defined as all sources of energy, except fossil fuels and nuclear energy. Nuclear energy is not renewable because it relies on non-replenishable, mined minerals such as uranium. Fossil fuels are also non-renewable because they take millions of years to replenish. The focus of this book is renewable energy, non-fossil and non-nuclear.

Renewable energy has been utilized for thousands of years in many parts of the world. For example, ancient Egyptians are among the earliest on record to use wind power to propel boats along the Nile River as far back as 5000 BC. Wind power has also been used for other purposes, such as windmills, in many parts of the ancient world. A windmill is a structure with rotating blades or sails that convert a blowing wind into rotational energy, which is used in grinding grain and pumping water. These wind power ingenuities of old times led to the wind energy technology of today.

Windmill used for grinding grain

Windmills use the power of the wind to spin the attached blades, which are connected to a drive shaft. The gear at the end of the drive shaft is connected to another gear and an additional shaft. At the bottom of the structure are two millstones. One is stationary while the other is attached to a spinning shaft. Grain is poured through a hole in the top millstone and grounded by the rotational force of the millstones. Finally, the flour exits on the side of the stone into flour sacs or silos below.

Vertical Blades

Earlier windmills were designed with vertical blades. They were designed to catch the blowing wind easily. However, with time, blades were designed and installed horizontally and attached to a central drive shaft for better efficiency.

Nashtifan, Iran

Another form of ancient renewable energy is the waterwheel, which is the oldest form of hydropower. Waterwheels use the kinetic energy of flowing water to spin attached paddles or buckets. The kinetic energy is then converted to rotational or mechanical energy and used for various purposes, including supplying running water for irrigation, grinding grain, and water pumps. Waterwheels were also used to supply energy for various industries of the day, such as sawmills, textile mills, and cast iron making. These ancient waterwheels are the pioneers of the modern-day hydroelectric dams which are used to generate nearly 7% of the United States and 15% of the global electricity generation today.

The earliest form of solar energy got its beginning thousands of years ago, as well. Over 3000 years ago, the Chinese used solar energy by creating concave reflectors to concentrate sunlight into more powerful rays for starting fires. Speaking of fire, heat from beneath the earth has been used in steam baths of ancient Chinese, Romans, and all over the world for healing and pleasure. The steam from beneath the earth is what is being developed as geothermal energy, increasingly used to generate global electricity supply. Finally, the earliest form of biofuel is animal manure dried up in the open field to be used as fuel for cooking or heating thousands of years ago.

Common types of Renewable Energy

There are many forms of renewable energy available today. Although all have been used for centuries, most are being developed for electricity generation to suit modern economies. These renewable sources include, Geothermal, Solar, Wind, Hydro, Bioenergy, and other organic and non-organic sources, such as algae and ocean waves.

Although many forms of renewable energy have been used for centuries, modern renewable energy began its journey in the early to mid-1800s with few scientific discoveries. Some of the findings include the work of Edmond Becquerel, a French physicist, who in 1839, discovered the effect of light and how it interacts with electrolytic cells. In the years to follow, many prominent scientists advanced the work. Honorary mentions include Professor William Grylls Adams of King's College in London and his colleague Richard Evans Day. In 1876, they demonstrated how certain materials, such as selenium, can produce an electric current when exposed to light. Fast forward to 1905, famed physicist Albert Einstein published a paper explaining the 'photoelectric effect, which is the emission of electrons when light is shined upon certain materials. A few years later, Albert Einstein won a Nobel Prize for Physics in 1921 for his work on the photoelectric effect, which is the driving force behind the modern-day solar energy industry.

Not to be outshined, the wind energy industry has a journey of its own. Besides the windmills and windpumps of yesteryears, wind turbines have been used for producing a small amount of electricity since the late 1800s. For example, Denmark, one of the most advanced countries in wind energy, started generating electricity from wind power as early as the 1890s. By 1908, Denmark had over 70 electricity-generating wind power systems.

In the western hemisphere, American settlers in the Western Great Plains were challenged with a shortage of water for their personal needs, for watering their livestock, and for growing crops despite the abundance of water deep underground. With the introduction of windmills by European migrants, they were able to resolve their water shortages by pumping water from great depths at a steady rate. Windmill pumps have the advantage of shifting into the prevailing winds. Thus, they are functional in both fast and slow currents. Most importantly, they require little maintenance. Soon, windmill pumps became a "must-have" for every homesteader, farmer, and rancher in the American West.

Windmill pump in a farm – Texas, USA 2021

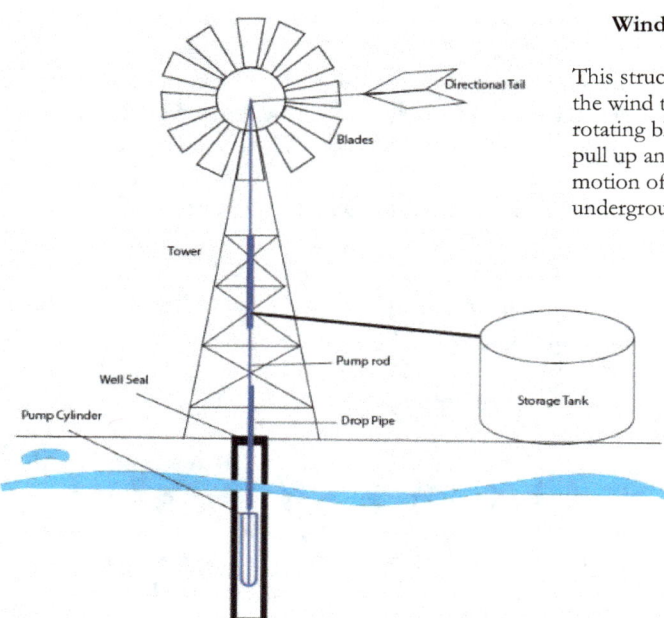

Windmill for pumping water

This structure also uses the power of the wind to spin attached blades. The rotating blades force an attached rod to pull up and down. The up and down motion of the rod helps pump underground water to a storage tank.

However, the wind energy industry matured in the United States following the oil shortages of the 1970s. Massive financial investment and favorable government policies helped the industry expand in the 1970s. By the 1980s, wind energy was integrated into the grid system of many utility companies to produce electricity for thousands of homes and businesses. Today, there are over 350,000 wind turbines in operation around the world generating electricity. According to the American Clean Power Association, the U.S. accounts for at least 60,000.

Other renewable sources, such as hydropower, developed from earlier technologies. Hydropower, for example, works similarly to the ancient waterwheels. A large amount of water flows down a tunnel into the turbine and spins the turbine blades the same way flowing water spins waterwheel blades or buckets. The rotating turbine spins an attached shaft, which turns the electricity-producing generator.

The Hoover Dam, located on the borders of Arizona and Nevada on the Colorado River is an example of modern hydropower achievement. The Hoover Dam project began in 1931 and at its completion in 1935, was the largest hydroelectric facility in the world with an electric generating capacity of 2.08 gigawatts (GW). According to the U.S. Bureau of Reclamation, the Hoover Dam generates 4 billion kWh of electricity annually for over a million homes in Nevada, Arizona, and California.

The reliability and affordability of hydropower have motivated many nations to invest in many large projects over the past few decades. Some of the most recent include the "Three Gorges Dam" in China. When completed in 2012, Three Gorges

| Water wheel of yesteryears used to re-direct water for irrigation | Hoover dam: a modern hydroelectric dam generating electricity for millions of homes |

hydropower became the largest project in the world with an electricity production capacity of 22.5 GW, nearly ten times the capacity of the Hoover Dam.

Thousands of miles further, on another continent, the "Grand Ethiopian Renaissance Dam" is one of the newest hydropower projects. The plant is still under construction with a planned capacity of 6.4 GW. When completed, the hydropower plant will be the largest in Africa, surpassing Egypt's Aswan Dam, which was completed in 1970 with a capacity of 2.2 GW.

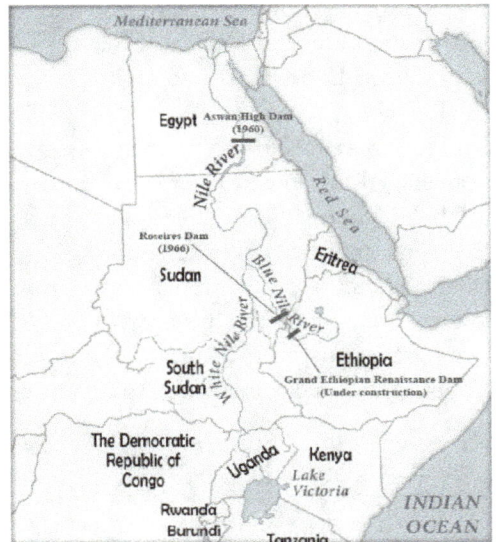

The newly constructed "Grand Ethiopian Renaissance Dam" on the Blue Nile

Finally, geothermal energy has a history as old as mankind. Although generating electricity with geothermal is a fairly new technology, using geothermal heat for bathing and heating is an old concept. For example, there are records of Native American tribes using natural steam spas for therapeutic purposes some 12,000 years ago. Similar records exist for cultures all over the world.

Accessibility

Renewable energy is limitless and everywhere. Where there is Mother Nature, there is renewable energy. The sun will always shine. The wind will always blow. Deep below the earth's surface, tectonic plates are continuously shifting, and volcanic activities are shaping our planet as they have for millennia. The unstable tectonic plates in southern California and the Great Rift Valley of East Africa provide geothermal energy for their regions. The Nile, the Congo, and the Columbia Rivers are great sources of hydropower, while the deserts of North Africa and the southwest United States can easily be turned into solar farms, giving life to large wastelands. Wind turbines are, of course, at home anywhere in the world.

Because renewable energy comes from a variety of sources, every corner of the world should benefit from at least one source. In most cases, there are multiple sources of renewable energy in a region that can be used as part of an energy portfolio mix. For example, the states of Texas and California take advantage of their sunny climate and vast desert areas for solar and wind energy farms. In the meantime, Washington State and Oregon benefit from hydropower projects built on the Columbia River that flow from the Canadian Rockies down to their territories. Globally, China has been the beneficiary of its vast and diverse landscape to generate a mix of hydro, solar, and wind energy for its booming economy.

Economically developing nations such as Ethiopia, Sudan, and Egypt benefit from the Nile River while neighboring Kenya has managed to generate over 40% of its electricity demand from geothermal energy. The geothermal energy Kenya enjoys comes from the East African Rift Valley which encompasses other neighboring countries, such as Eritrea, Uganda, and Rwanda. Therefore, there remains untapped geothermal energy in the region. To its credit, the state of Eritrea has been conducting multiple studies to analyze the energy potential of its Danakil region on the coast of the Red Sea and follow in the footsteps of an already successful Kenya.

Energy prices

Most industrialized nations are spoiled with cheap energy prices from fossil fuel and nuclear energy-driven utilities. For example, in the U.S., the cost of electricity for residential, commercial, and industrial are about 13 cents, 11 cents, and 7 cents per kWh respectively. Fossil-fuel-based power plants account for 60% of the electricity generated while nuclear energy-fueled plants contributed an additional 18% to the overall U.S. electricity production in 2022. Unfortunately, the cheap energy price is one of the obstacles to change. For the renewable energy industry to be considered an alternative to fossil fuels, it needs to provide electricity at a competitive cost.

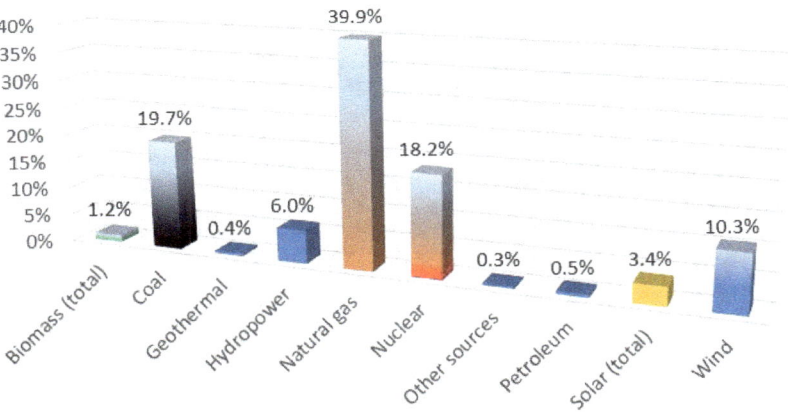

Non-renewable energy, including nuclear, contributed to 78% of U.S. electricity production in 2022

Fortunately, the renewable energy industry has progressed over the last few decades. Recent data shows the industry can compete with traditional fossil fuel-based plants and generate electricity at a competitive cost. In fact, the cost of electricity from renewable energy-fueled plants is the cheapest in the market today. The below chart shows pricing trends in the U.S. electricity market over the last 30 years. For example, the price of generating electricity with solar power has decreased from 38 cents per kWh in 2010 to 3.9 cents in 2021, making it the cheapest source of electricity. It is an amazing technological accomplishment to witness the price of electricity from solar reduced by nearly 90% in a short 10 years. Onshore wind power is the second cheapest source of electricity at 4.3 cents per kWh. Other sources, such as hydro, are not far behind at around 5 cents per kWh. Meanwhile, electricity generation from Geothermal sources is increasingly becoming widespread. Geothermal sources offer one of the cheapest sources of electricity as well.

Globally, similar trends are observed. The cost of electricity from both wind and solar-based plants has decreased from north of 30 cents per kWh to below 10 cents per kWh. Below is a pricing trend for the U.S. market and the markets of 19 other industrialized nations.

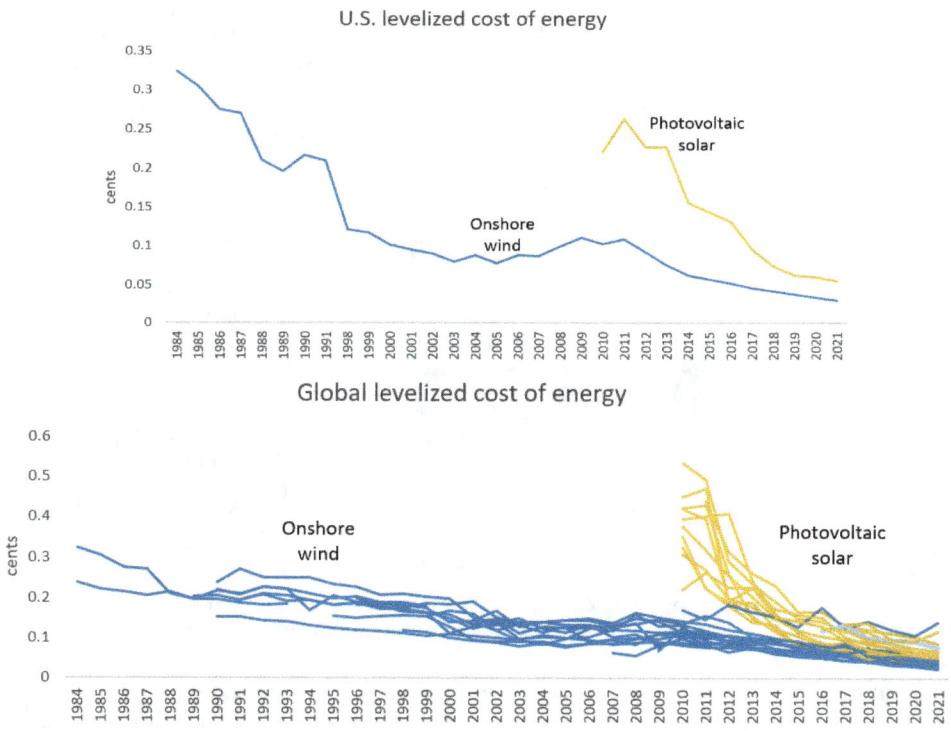

Employment

Access to multiple sources of renewable energy is a start, but transforming these sources into an economic engine can prove more challenging. During ancient times, renewable energy was directly utilized to do work. Water and wind were directly used as mechanical energy. Waterwheels used water from streams for grain grinding, just as were windmills. Geothermal energy was directly used to heat space or for steam bathing. In today's modern economies, however, energy from all renewable sources must be converted into electrical energy or electricity. Other forms, such as biofuel must be converted to mechanical energy for our vehicles, while other forms of energy are needed for heating and cooling. Energy conversion requires technological advancement that only a few nations possess and financial investments only a few nations can afford. To those with the know-how and financial means to transform renewable energy into useful energy, employment opportunities are the reward. To those who lack the means, renewable energy remains untapped economic potential.

Below is data from the U.S. Bureau of Labor Statistics on the U.S. job market and the energy/utility sector's contribution to the job market. According to the data, the U.S. market created over 164 million jobs in 2022 with 98.6% of the jobs in non-agricultural sectors. As shown in the table below, the most prominent sectors are government, healthcare, retail, manufacturing, and services. The top 5 industries, which do not include the energy sector, make up nearly 50% of all jobs in the U.S.

Employment by major industry sector (Employment in thousands of jobs)
Source: U.S. Bureau of Labor Statistics

Industry Sector	Employment, 2012	Employment, 2022	Employment, 2032	Percent distribution, 2012	Percent distribution, 2022	Percent distribution, 2032
Nonagriculture wage and salary	134,844.2	153,185.5	157,779.4	92.6	93.1	93.3
Agriculture, forestry, fishing, and hunting	2,084.3	2,184.4	2,126.9	1.4	1.3	1.3
Nonagriculture self-employed	8,733.5	9,112.6	9,241.8	6.0	5.5	5.5
All industries	**145,662.0**	**164,482.6**	**169,148.1**	**100.0**	**100.0**	**100.0**
Nonagriculture wage and salary	134,844.2	153,185.5	157,779.4	92.6	93.1	93.3
Goods-producing excluding agriculture	18,370.1	21,133.6	21,120.6	12.6	12.8	12.5
Mining	797.2	559.9	545.4	0.5	0.3	0.3
Construction	5,646.0	7,748.0	7,862.9	3.9	4.7	4.6
Manufacturing	11,926.9	12,825.7	12,712.3	8.2	7.8	7.5
Service-providing excluding special industries	116,474.1	132,051.9	136,658.8	80.0	80.3	80.8
Utilities	552.8	553.6	539.2	0.4	0.3	0.3
Wholesale trade	5,595.2	5,962.6	5,877.8	3.8	3.6	3.5
Retail trade	14,800.9	15,475.4	14,946.3	10.2	9.4	8.8
Transportation and warehousing	4,403.8	6,651.1	7,221.0	3.0	4.0	4.3
Information	2,676.0	3,074.4	3,275.9	1.8	1.9	1.9
Financial activities	7,783.4	9,044.5	9,393.8	5.3	5.5	5.6
Professional and business services	18,037.0	22,571.5	23,991.7	12.4	13.7	14.2
Education services	3,341.0	3,794.7	3,924.3	2.3	2.3	2.3
Health care and social assistance	17,428.0	20,555.0	22,647.0	12.0	12.5	13.4
Leisure and hospitality	13,768.2	15,835.2	16,185.8	9.5	9.6	9.6
Other services	6,167.7	6,362.7	6,463.5	4.2	3.9	3.8
Federal government	2,820.5	2,869.4	2,825.5	1.9	1.7	1.7
State and local government	19,099.6	19,301.8	19,367.1	13.1	11.7	11.4
Agriculture, forestry, fishing, and hunting	2,084.3	2,184.4	2,126.9	1.4	1.3	1.3
Agriculture, forestry, fishing, and hunting wage and salary	1,307.3	1,441.6	1,454.2	0.9	0.9	0.9
Agriculture, forestry, fishing, and hunting self-employed	777.0	742.8	672.6	0.5	0.5	0.4
Nonagriculture self-employed	**8,733.5**	**9,112.6**	**9,241.8**	**6.0**	**5.5**	**5.5**

According to US Energy & Employment Report 2022, the energy sector contributed nearly 8 million jobs or 4.5% to the overall U.S. job market in 2022. The jobs are dispersed in various sectors including utilities, construction, services, and manufacturing (including the manufacturing of Electric Vehicles/EVs). Below is a breakdown of employment in the energy sector.

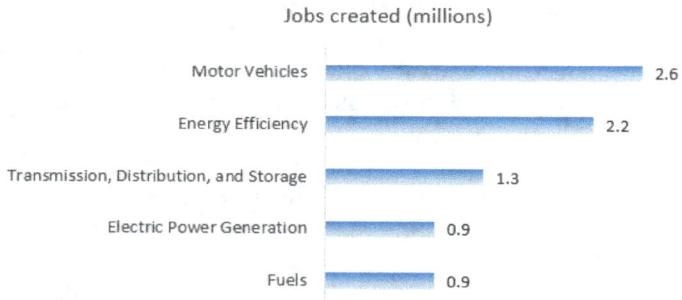

Motor Vehicles: Motor vehicles and parts. This sector includes carbon-reducing motor vehicles, such as hybrid and electric vehicles, and parts technologies.

Energy Efficiency: The energy efficiency sector is also another promising sector where over 2.2 million jobs are created. This sector creates employment in the production and installation of energy-efficient products to help reduce energy consumption. Traditional heating, ventilation, and air conditioning (HVAC), energy-star appliances, and building efficiency are some of the examples. The Energy Star program alone creates jobs in energy-saving appliances and LED light manufacturing. In addition, this sector employs many in the construction, manufacturing, and installation of equipment, such as solar panels, wind turbines, and generators.

Transmission, Distribution, and Storage (TDS): This sector includes traditional transmission and distribution as well as smart grids. It is also becoming more apparent that the reliability of solar and wind power will heavily depend on battery storage technology. As a result, this sector has gained recognition as battery technology is finally coming to light for utilities and EVs.

Electric Power Generation: Includes electricity generated using renewable sources, such as solar, wind, and hydro as well as electricity generated using fossil fuel and nuclear energy. Renewable energy sources contribute to 63% of all jobs in this sector with solar and wind topping all. This sector also includes employment in firms engaged in the manufacture, operation, and/or maintenance of turbines and other generating equipment. In addition, employment in the construction and installation of electricity generation plants, capital investments, and wholesale parts distribution for all electric generation technologies are included in this sector. Most importantly, since most of these jobs are in construction and utilities, they will remain local and contribute to the job market of local economies.

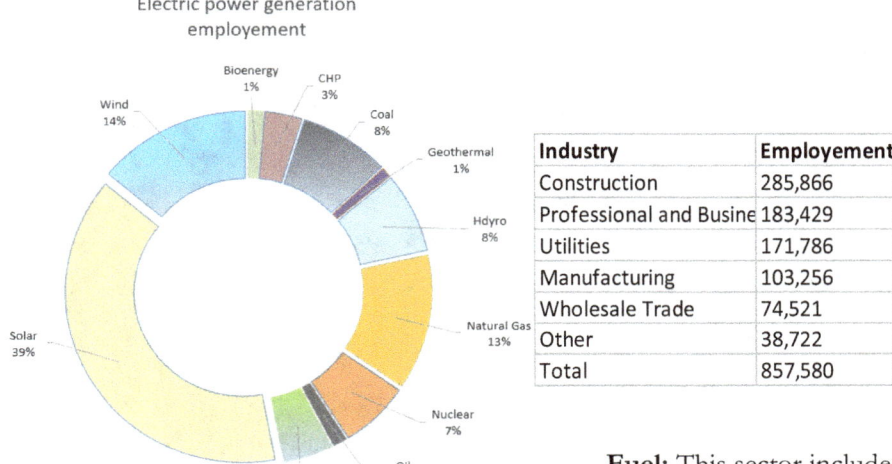

Industry	Employment
Construction	285,866
Professional and Busine	183,429
Utilities	171,786
Manufacturing	103,256
Wholesale Trade	74,521
Other	38,722
Total	857,580

Fuel: This sector includes both fossil fuel (coal, petroleum, and natural gas) and fuel from renewable sources of biofuel, including renewable diesel fuel, biodiesel fuel, and waste fuel. In addition, the transportation sector has made great strides in converting from fossil fuels to renewables. Thus, biofuel and biodiesel have become just as common, while Electric Vehicles (EVs) have been on the rise reducing the need for petroleum.

The energy sector makes up a small portion of the U.S. job market. The sector contributes nearly 8 million jobs and accounts for 4.5% of the overall 164 million jobs in the country. However, a closer look at the energy labor market shines a light on the promise of the renewable energy industry. Over 3.2 million jobs or 43% of the 8 million energy jobs are in the renewable energy industry. This is astonishing considering the industry is still in its infancy. As the industry develops, there is no doubt it will continue to provide a cleaner and more lucrative source of employment and livelihood for many.

Globally, the energy sector employed over 65 million workers accounting for 2% of the global workforce in 2019, according to the IEA report published in 2022. As the table below illustrates, the ratio is evenly divided between fuel supply, power generation, and end users which includes efficiency. The "energy efficiency" and the "power generation" sectors account for 33.8% of energy sector jobs. Most of the jobs in these two sectors are created by the renewable energy industry.

Employment by region and energy sector in thousands of employees, 2019

	North America	Central and South America	Europe	Africa	China	India	Other Asia Pacific	Rest of world	Global
Supply: coal	100	<50	100	200	3 400	1 400	800	300	**6 300**
Supply: oil and gas	1 900	1 100	600	1 600	1 100	700	1 100	3 800	**11 800**
Supply: bioenergy	100	800	300	600	300	500	600	<50	**3 300**
Power: generation	1 000	600	1 400	400	3 800	1 200	1 800	1 000	**11 300**
Power: grids	900	400	1 200	500	2 300	1 500	1 200	600	**8 500**
End uses: vehicle	1 800	600	2 700	200	4 500	1 200	2 100	600	**13 600**
End uses: efficiency	2 000	300	1 100	400	3 800	1 500	1 400	400	**10 900**
All energy	7 900	3 800	7 500	3 800	19 200	7 900	8 900	6 600	**65 700**

Self-Sufficiency

Economies are dependent on globalization more today than ever before. Resources and production are specialized in certain regions and transported across the globe for consumption. Therefore, international conflicts and instability can easily disrupt economic activities, such as transnational freighting. For example, in the recent Russia-Ukraine conflict, not only military hardware, but finance, freight, and economic diplomacy are weaponized. The supply of wheat and fertilizers from Russia and Ukraine to the global markets is disrupted. Cheap natural gas from Russia to the European markets has halted, impacting European economies. Pipelines used to transport natural gas from Russia to Western Europe, particularly Germany, are mysteriously blown up. Furthermore, regional alliances, such as OPEC, continue to amass greater influence on oil and gas pricing and supplies to the global market.

The best way nations can protect their economies from market disruptions is to achieve energy self-sufficiency. Self-sufficiency, however, is a two-step process. First, there must be energy accessibility. As previously mentioned, every nation, if not most, has access to one or multiple sources of renewable energy. The second step is to convert the locally available sources into usable energy. Flowing rivers need dams, generators, turbines, and other equipment to generate hydroelectric power. The wind blows more consistently at a higher elevation. Thus, wind power needs large blades and generators on tall towers to produce electricity efficiently.

Sunshine must be captured by solar panels. The manufacturing of the wafers that make up solar panels is highly sophisticated and sensitive. All the hardware and software required to turn the energy resources into usable energy requires technological advancement and financial commitment. If nations can master both the technological and financial means, then they can achieve self-sufficiency.

For example, China manufactures most of the solar panels in the world. In fact, according to 2021 data from the International Energy Agency (IEA), China manufactured 75% of all solar panels in the world in 2021. In addition, China manufactured 85% of the cells, 97% of the wafers, and 79% of the polysilicon, dominating not only the solar panel production but also the components that go into the panels. When it comes to wind energy, the manufacturing of wind turbines, nacelles, and other components is diversified. While China still makes up over 60% of the manufacturing capability, European and U.S. companies have a significant manufacturing capability as well.

Other nations, particularly developing nations, have the natural resources but lack the technological and financial means to develop them. For example, Ethiopia and Sudan are blessed with water resources from the Nile. Hydroelectric power from the Nile can easily fulfill a hundred percent of their electricity demand. Unfortunately, the water turbines and generators required for the hydroelectric plants must come from abroad. The knowledge and finance to build the hydroelectric dams come from abroad as well. Thus, these nations cannot be considered self-sufficient despite the availability of limitless energy sources.

Besides technological and financial challenges, international politics can also be an obstacle to achieving energy self-sufficiency. For example, trade barriers or sanctions exercised by many Western nations towards other non-Western nations have halted many renewable energy projects, especially in the global south. These political meddling by the West are designed to preserve colonial-era policies and maintain a grip on former colonies which have been a supply of natural resources to the world for centuries. For example, the U.S. and European-led sanctions imposed on Eritrea have prevented the nation from becoming a hundred percent energy self-sufficient.

Located on the coast of the Red Sea in the horn of Africa, Eritrea is politically a young, but historically an old nation. The country has a history as old as the history of mankind itself. It is scientifically believed that the first humans migrated out of Africa to Asia & the rest of the world by crossing this region of Africa. The belief is supported by the many ancient artifacts found in the coastal towns of Eritrea. Eritrea also has a rich history of being part of the past African kingdoms of Punt, Kush, and Axum stretching from 5000 B.C. to 1400 A.D. Politically, however, this nation is relatively new on the world stage, having won its independence in 1991 after decades-long war.

Shortly after independence, the government of Eritrea partnered with many experts to develop its energy sector, which is key to its economic advancement. After all, the country is blessed with natural energy resources. For example, Eritrea's 1100km (680 mi) coastline has proven petroleum and natural gas resources. Its arid and hot lowlands have great potential for solar energy. The highlands offer an excellent opportunity for wind energy, while the Danakil region is a great source of geothermal energy.

In 1996, the Eritrean government partnered with the United States Agency for International Development (USAID) and the United States Geological Survey (USGS) to exploit its geothermal resources in the Danakil depression region. The Danakil Depression is a vast plain, 200 km by 50 km (124 mi by 31 mi), lying in the north of the Afar Region of Eritrea and Ethiopia. It is about 125 meters (410 ft) below sea level, making it the shallowest place on earth. Geologically, the Danakil Depression lies at the triple junction of three tectonic plates; the African, Arabian, and Indian tectonic plates. In geological terms, it is the result of Africa and Asia moving apart, causing rifting and volcanic activity.

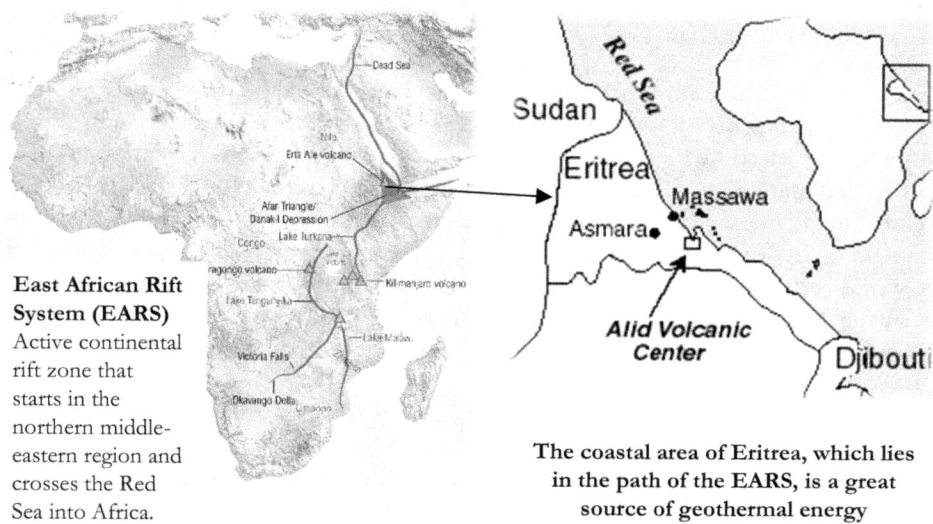

East African Rift System (EARS) Active continental rift zone that starts in the northern middle-eastern region and crosses the Red Sea into Africa.

The coastal area of Eritrea, which lies in the path of the EARS, is a great source of geothermal energy

The study was more focused on the Alid volcanic center. Extensive studies have taken place to produce a geothermal resource assessment based on mapping of the geology and study of the fumarolic emissions from the Alid volcanic center. The study concluded that

> "compositions of fumarolic gases collected at Alid indicate that the reservoir temperature of a hydrothermal-convection system driven by this heat source is very likely in the range of 250° to 300°C. The overall temperature and permeability conditions seem so favorable for an electrical grade geothermal resource that exploration drilling to depths of 1.5 to 2 km is recommended".
> source: https://pubs.usgs.gov/of/1997/0291/report.pdf

Unfortunately, all efforts of energy development by the Eritrean government were kneecapped by the trade sanction imposed by the United States and the European Union in the late 1990s. Since advanced equipment, such as turbines and generators, has to be imported, the country's geothermal energy project along with economic development remains halted.

Sulfuric geothermal acid lake in the Danakil

The renewable energy industry can contribute to energy self-sufficiency for many nations. Renewable energy is more accessible and more affordable than fossil fuel. It is also available in abundance and variety. Renewable sources-driven power plants generate cheaper electricity than fossil fuel-based plants. As previously stated, solar power and wind power are the cheapest sources of electricity. The job market is also promising. Even at its earliest stages, the industry accounts for nearly half of the employment in the energy sector job market. Therefore, the advantages of renewable energy must be communicated clearly so that a shift towards renewables is voluntary. Climate change fear-mongering cannot be the tool of persuasion. People must be convinced of the economic opportunities that can be achieved by embracing renewable energy. After all, there is an opportunity to support both sides of the climate change and the energy debate. An opportunity to shift towards cleaner, cheaper, and renewable energy sources while contributing to a cleaner environment.

Energy Efficiency

The United States has allocated nearly $1 trillion in 2023 to bolster its renewable energy economy. The budget known as the Inflation Reduction Act (IRA), is meant to energize the U.S. renewable energy industry by investing in energy supply chains, emissions reduction, consumer energy savings, and clean energy jobs. The impact of the investment is unknown and will take time to analyze. However, similar investments in past years have played a significant role in developing the renewable energy industry in the United States. One of the sectors that benefited most from the investments is the energy efficiency sector. After all, the energy we don't use is the energy we don't have to generate.

Previously, the United States had invested $1.8 Trillion (USD) in the energy sector in 2018. $240 billion, nearly 13%, was allocated to the energy efficiency sector. The sector is classified into three categories: Building efficiency, Transportation efficiency, and Industrial efficiency. The building sector, which includes efficiency for the heating & cooling of buildings in both residential and commercial, consumed a significant portion of the budget, nearing $140 billion, according to the International Energy Agency (IEA) 2019 report. An additional $60 billion was allocated for the transportation industry, including heavy-duty and light-duty freight and electric vehicles (EV). The remaining $40 billion was budgeted for efficiency improvement in the industrial sector.

The massive investment has helped the energy efficiency sector create over 2.2 million jobs in the United States as of 2021. The sector includes ENERGY STAR appliances and lighting, pollution reduction and removal, greenhouse gas reduction, recycling, natural resources conservation and environmental compliance, and retail selling of energy-efficient products. The jobs are dispersed across the country, with workers employed in all but seven counties, according to the Environmental and Energy Study Institute (EESI). More than 300,000 people are employed in energy efficiency in rural areas and 900,000 people work in energy efficiency in the country's 25 largest metro areas.

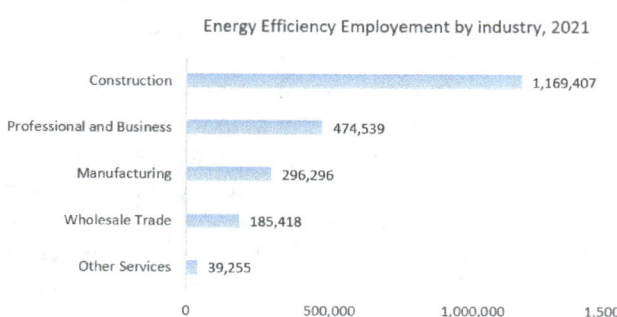

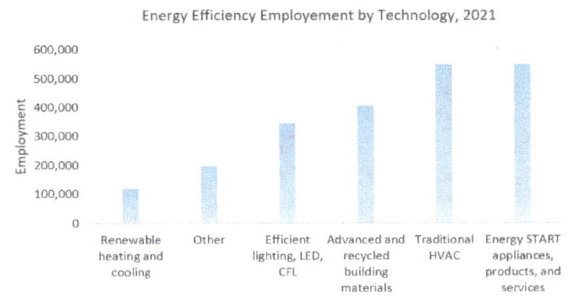

Additionally, according to the U.S. Energy and Employment report, over 800,000 Americans are employed in manufacturing or installing ENERGY STAR-certified appliances, including heating and cooling.

Buildings efficiency: time-tested techniques

As previously mentioned, a significant portion of the energy efficiency budget was allocated to the building sector because of the intense energy required for heating and cooling. The investment was designed to incentivize energy efficiency in the heating, cooling, and powering of residential, business, and industrial facilities. The technological advancements in construction techniques and materials that followed, such as lightweight-efficient insulations, can be attributed to these investment decisions. The investment has also contributed to the manufacturing of efficient modern appliances, such as Energy Star-rated appliances.

Equally, efforts are made to reevaluate past techniques and experiences that may have been overlooked. Modern heating and cooling appliances are a recent phenomenon. For millennia, mankind has found ways to create a comfortable living space using locally available materials and time-tested techniques, including selecting the right construction material and building with nature in mind. These ancient and cost-effective techniques, known as Passive Solar Design, are being incorporated into modern architecture to help achieve efficiency. Passive Solar Design is categorized into passive solar heating, which is a method used to heat living spaces, and passive solar cooling, which is a method used to cool living spaces. Passive solar design is achieved using orientation and material selection.

Passive solar heating: orientation

Passive solar heating is the concept of using the sun's energy to heat a living space and is utilized during colder fall or winter seasons. The idea is to expose the building to as much of the sun's light and heat as possible during the day. This is accomplished by orientating openings such as windows, doors, and sunroofs to face the sun. In the northern hemisphere, such as North America and Europe, buildings would face true south. In contrast, in the southern hemisphere like Africa, South America, and Australia, buildings would be designed to face true north. True south and true north are where the sun is at its highest. True north/south is different from commonly used directions of north/south. It is important that we understand the difference between the two.

The earth contains a giant magnet. When we use a directional compass, the compass needle which is a tiny magnet, lines up with the poles of that giant magnet. This is called south or north or magnetic south or magnetic north. However, since the magnet is not aligned with the earth's rotation, the compass needle does not line up with the poles that the earth rotates around. Therefore, we must account for the earth's rotation to find the true north or true south where the sun's energy is maximized.

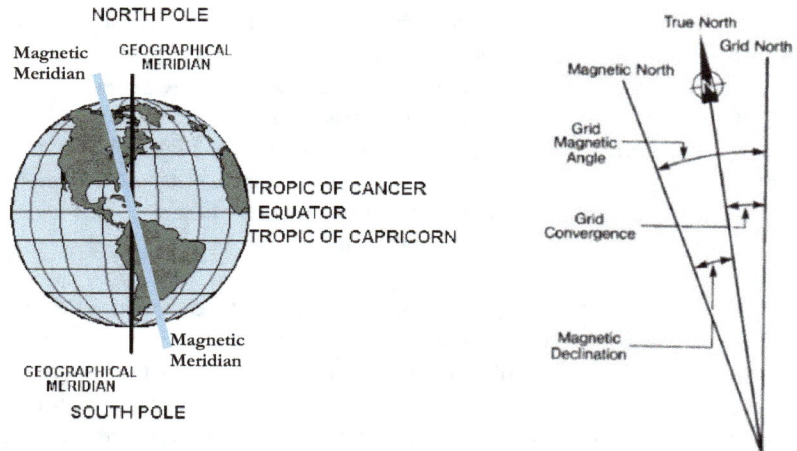

True south/north versus magnetic south/north

In addition to orientation, understanding the energy radiated from the sun to our planet is beneficial to maximizing the energy gain that can be achieved. The average temperature of the sun at the surface is believed to be 10,000 °F/5500 °C. Most importantly, the amount of energy the sun radiates to earth in one hour is equal to the amount of energy we need in an entire year. It would be unwise not to harness this abundance of free energy. The energy radiated from the sun is classified according to its wavelength, ranging from long wavelength (weak energy) to short wavelength (high energy). While nearly half of the radiated energy is considered visible light, the other half is considered infrared or heat energy. Harnessing the energy of the visible light requires solar panels and other mechanical devices in what is known as "active solar energy". Active solar energy will be discussed in detail in later chapters. Passive solar heating, on the other hand, takes advantage of the non-visible, radiated infrared or heat energy.

The sun's energy is radiated through space and into the earth's atmosphere where it is disbursed in many directions. About 35% to 40% of the energy is reflected to space from the clouds and atmospheric dust before it reaches earth. Ultra-violet high frequency or high energy radiation, which can cause skin burns, skin cancer, and eye damage, is part of the radiation that is reflected to space and does not reach the earth's surface. The ozone layer, which is in the upper atmosphere, plays an important role in blocking this ultra-violet high energy radiation from reaching the earth's surface. Thus, the ozone layer plays a crucial role in protecting humans from the dangerous parts of the sun's energy. In recent years, a depleting ozone layer has been a critical argument for climate change proponents. There is a belief that man-made atmospheric pollution is depleting or thinning out the same ozone layer that protects humanity from harmful radiation.

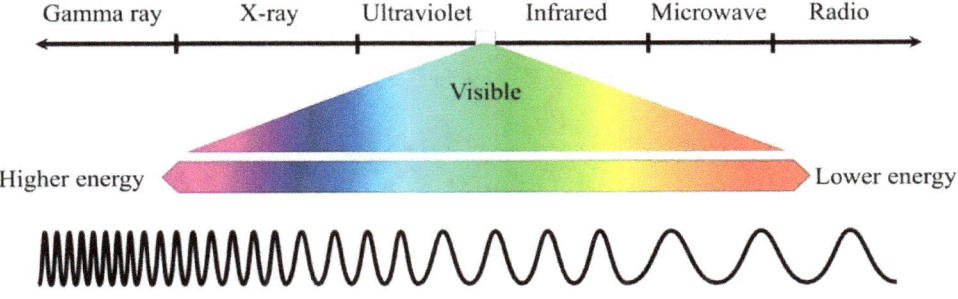

In addition, some portion of the sun's energy that penetrates through the earth's atmosphere is reflected to space from the surface of water/oceans, sand, and snow. Overall, nearly half of the sun's energy that is radiated to the earth's atmosphere is unavailable for passive solar heating because it is reflected to space. Furthermore, some portion of the sun's energy interacts with air molecules and dust particles in our atmosphere, and rather than reflected, is scattered, or diffused throughout the atmosphere. The scattered energy/radiation is primarily the blue portion of the visible spectrum which is responsible for the blue color of the sky. Again, since this portion of the radiated energy does not reach the earth's surface, it also is unavailable for solar heating. However, during cloudy days, when direct radiation from the sun is minimized by clouds, almost all the energy that penetrates the earth's surface comes from some of the diffused and scattered radiation. Hence, a small amount of passive solar heat is available even on cloudy days.

In addition to the makeup of the atmospheric layers and the composition of the radiated energy, the length and angle the radiation must travel are key determinants of how much solar radiation reaches the earth's surface. The length and the angle of the radiated energy are impacted by the earth's axis rotation and tilt. For example, during summer middays when the sun's path is directly overhead, the sun's rays are nearly perpendicular to the earth's surface and the radiation travels through the least amount of atmosphere en route to the earth's surface. Thus, the energy is very intense. At sunset, however, the path lengthens, and the perpendicularity of the rays lessens. As a result, the energy intensity weakens.

The more atmospheric mass the radiation travels through, and the less perpendicular its path, the lower the energy intensity. This is due to increased absorption and scattering of radiation. Thus, atmospheric air mass and angularity determine the intensity of the direct radiation. The length of travel and the intensity of the radiation varies with the time of day and month of the year due to the earth's tilt and rotation. Since the tilt is consistent as the earth orbits around the sun, places around the equator receive long hours of sunshine during the day and throughout the year. Therefore, regions along the equator and a few degrees north and south of the equator benefit the most from passive solar heating.

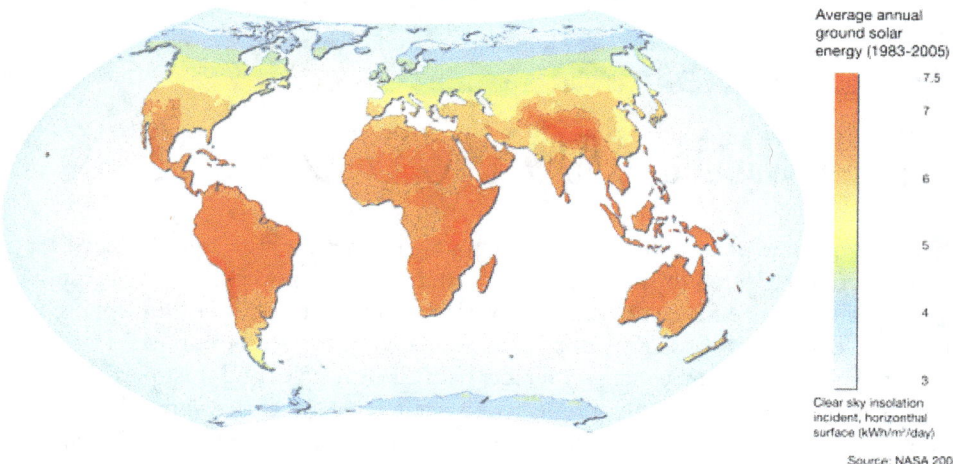

Fortunately, most of the sun's energy penetrates the earth's atmosphere and is intercepted by the earth's surface. This is called direct radiation and includes all the high-energy UVs, low-energy UVs, and the visible lights of the energy spectrum.

Direct radiation is the building block of all life on earth. It is also the source of solar passive heating. With the understanding of how the sun's energy is radiated to our planet and how the earth's rotation and tilt affect the radiated energy, we can plan for passive solar heating more effectively. Additionally, the understanding of the expected energy intensity during the day, throughout the year, and at different seasons, is crucial for the utilization and optimization of the sun's energy towards an effective passive solar design. In fact, homes re-oriented toward the sun without any additional solar features save between 10% and 20% and some can save up to 40% on home heating, according to a study by the Bonneville Power Administration and the City of San Jose, California.

Passive solar cooling: material selection

During the summer seasons, the sun's energy is intense and thus, we need a mechanism to shelter from its scorching heat. The term passive solar cooling refers to the scenario where our efforts are concentrated on blocking the sun's energy. It is the opposite of passive solar heating. There are simple and effective ways of accomplishing the task. For example, since the sun's path follows a high arc during summer, roof overhangs are very effective in blocking the sun's rays. Additionally, the following simpler and time-tested tools are all effective:

- blinders provide more control over how much sun can penetrate our spaces.
- Trees that blossom during summer (thus blocking the sun's rays) and shed their leaves during winter (thus allowing more sunshine to pass through).
- Glazed windows, which are a more technologically advanced tool, to help minimize sun exposure.

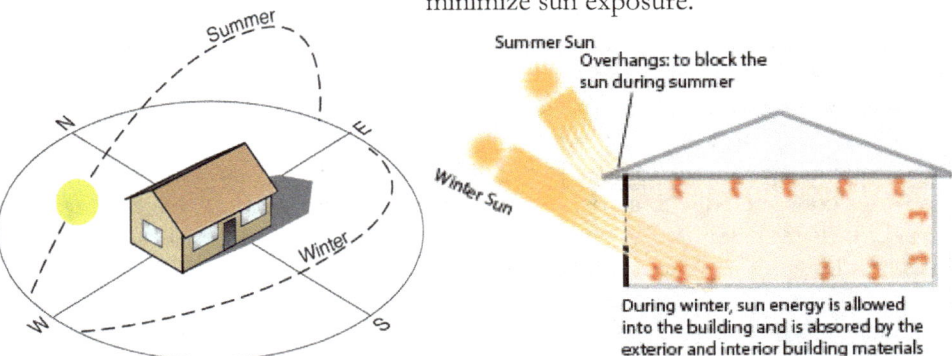

The sun's path varies throughout the year.
It follows higher arc during summer compared to winter

However, the most passively effective way of providing a cool living space during hot summer days is by using the right building materials during new construction or when modifying existing buildings. Certain building materials can absorb the sun's energy for a longer period before transferring the heat to the interior. Therefore, incorporating these materials in our building structures helps keep the inside space cool and comfortable. These materials do not block or reflect the radiation but absorb it instead. The absorbed energy is eventually released into the building interior hours later when room temperatures have dropped. Material selection is widely used in climates where there is a sizeable temperature swing between day and night. The right material helps regulate temperature fluctuation and contributes to a more comfortable living space.

The preferred material is one that can absorb energy for longer periods. The selection is accomplished by analyzing the material's thermal mass. Thermal mass is a material's capacity to absorb energy. Thus, building materials used in passive solar energy design are ranked according to their thermal mass. Those with higher thermal mass are more preferred. The material's thermal mass is formulated from its heat capacity (how much heat it absorbs) and density (mass or volume). Heat capacity is defined as the amount of heat energy required to change the temperature of an object. For example, it takes a lot of heat energy to change the temperature of concrete or clay compared to metal. If we heat a block of metal and a block of clay brick on a stovetop for the same amount of time and then try to pick them up, we are more likely to get a skin burn from the metallic block. This is because metal transfers the heat from the stovetop to our skin immediately. However, we do not feel the same heat intensity from the clay brick because the heat is absorbed by the brick rather than transferred to our skin. The heat is eventually released or transferred from the brick to our skin gradually over a longer period. The gradual heat release gives us a warm feeling and not a burn. Thus, we can conclude that clay has a higher heat capacity than metal and is more desirable for passive solar design.

Another example is materials used for rooftops. Rooftops built with metal sheets (commonly seen on warehouses or industrial structures) will transfer all the sun's heat immediately into the interior space and create a hot and uncomfortable space. Rooftops constructed with earthy materials, on the other hand, will have a much cooler inside temperature because the heat will be absorbed by the roof material. This is a design more commonly seen in homes built in a desert climate.

In addition to heat capacity, the total amount of energy an object can absorb depends on its density or mass. Therefore, high-density materials like tile, concrete, or brick absorb much more energy and are more desirable for passive solar design.

The table below shows a list of materials and their thermal mass rankings (Effective heat Capacity). The higher the ranking, the more desirable the material is for passive solar design.

Building Material Material	Specific heat Capacity (J/kgK)	Density (kg/m3)	Effective heat Capacity (Wh/m2K)
Water	4200	1000	175
Cast concrete	1000	2000	83.3
Concrecte Block	840	2240	73.1
Brick	800	1750	42.4
Timber	1600	650	5.4
Ceramic Tiles	800	1900	4.2
Wet Plaster	1000	1330	3.7
Plasterboard	840	950	2.7
Stone	900	2000	
Earth Wall (Adobe)	837	1550	

Water has the highest heat capacity. This makes water the most effective material for passive solar design. However, it is quite challenging to construct a building with water.

In some instances, people are experimenting with water filled containers or water barrels to utilize water's heat absorption capabilities. Although not suitable for building, it can serve a purpose for smaller projects.

Commonly used building materials with high thermal mass include:

- Concrete, clay bricks, and other forms of masonry: Concrete's thermal mass properties save 5-8% in annual energy costs compared to softwood lumber.
- Earth, mud, and sod: In addition to using these materials for building, people sometimes use earth sheltering around their homes for the same effect. Basements are a great example. They are dug in below ground and thus sheltered by earthy material.
- Logs or solid wood have significant thermal mass like concrete or earthy materials. In addition, solid wood has the added benefit of a much better insulation characteristic. For these reasons, log homes are common in colder climates. The insulation factor keeps the inside temperature insulated and warm during winter times, but the thermal aspect absorbs much of the daytime heat and helps keep the interior space cooler during summer.

The Native Americans' cliff dwelling (Pueblo ruins) in Mesa Verde, Colorado, southwestern part of the U.S. is a perfect example of utilizing location, orientation, and material to create a comfortable interior space in an arid and hot environment. These combinations help minimize the interior heat during the daytime and create warmer interior temperatures at night. It is believed that the dwelling was constructed in the late 1190s and was home to a native American tribe for nearly a century. According to the National Park Service, the structures ranged in size from one-room granaries/grain storage to villages of more than 150 rooms. These structures were built on the side of a cliff to minimize direct sun exposure. The

Native American Cliff Dwelling Mesa Verde, Colorado

earthy materials also absorb much of the heat during the day. The absorbed heat is gradually released into the interior when night temperatures are much colder. Since this part of the United States experiences sizeable temperature fluctuation between day and night, the concept of passive solar design works perfectly here. The main purpose for the cliff dwelling may have been for defensive or other purposes. However, knowledge inherited from past generations in addition to knowledge gained through experience and trial and error has allowed the builders to incorporate orientation and material selection into their structures. After all, the concept of passive solar design is as old as mankind.

In other parts of the world, people have also used similar techniques of passive solar concepts. For example, in parts of Eritrea, east Africa, traditional homes, known as Hidmos, have been built using earthy materials for centuries. The walls are built with stones while roofs are constructed of logs strengthened with a mixture of hay straws and earth. The chosen earthy material makes Hidmos strong and long-lasting structures. Hidmos that were built hundreds of years ago are still standing today. The mixture of earthy materials also makes them a model for passive solar energy enthusiasts. In this part of the world where the fluctuation between day and night temperatures creates uncomfortable living conditions, Hidmos have proven to be a very cool place during the day and warm and comfortable at night.

Hidmo in the Eritrean highlands. The masonry walls and a roof made from logs, hay straw, and earth makes a great passive solar energy design.

In other regions, homes were built partially underground or under earthy materials to exploit the minimal temperature changes below ground. The sunken houses of the Japanese and the pit houses of the Vikings are some examples. Our ancestors did not have the luxury of modern science to translate their work in terms of specific heat and heat capacity. However, they have perfected passive solar design through time, experience, and trial and error. Their work is slowly becoming a foundation and a template for many modern-day architects.

Traditional home in Ireland. A combination of masonry walls and roof made from logs and hay straw, demonstrates a perfect passive solar energy concept.

Today, there is an abundant choice of construction materials in addition to the time-tested materials of the past. Unfortunately, energy conservation is not considered when building. Modern buildings are designed and built for accommodation, fashion, scenery, and convenience. For example, skyscrapers made from steel and glass tower city landscapes all over the world. The steel is necessary for strength while the glass provides amazing scenery. Indoor shopping malls are designed for convenience while the construction of residential buildings prioritizes comfort. The lack of consideration for energy conservation has been possible because of advanced heating and cooling systems (HVAC) and the affordability to keep them running.

In recent years, however, the lack of planning has become costly. HVACs consume a tremendous amount of electricity. Considering the rising cost of electricity, maintaining these modern structures has become expensive and unaffordable. In fact, in many parts of the Western world, a third of living expenses is allocated for heating and cooling purposes. Forced by the rising cost and climate change awareness, many architects and home builders are incorporating energy-saving mechanisms into their designs. They are finally realizing the importance of utilizing ancient techniques alongside modern ways. Time-tested methods of orientation and material selection are increasingly incorporated into modern home designs to reduce energy consumption and energy expenses.

Today, home builders are embracing earthy materials such as stone, stucco, bricks, and concrete blocks for siding. The material used for the interior of the building is just as equally important. When the sun's energy enters a living space through doors or windows, the energy strikes the floor at an angle. Depending on the floor material, the energy will either be absorbed or disbursed into the home. Most homes in the Western world use carpet flooring. Carpet or fabric has one of the worst thermal mass and thus, is undesirable for passive solar cooling. Wood or preferably concrete or clay flooring tiles are the preferred choices. Concrete or clay tiles have an excellent thermal mass and can absorb the heat that enters the home. In other parts of the world, where earthy materials are not used, the chosen material is carefully considered to provide the same "temperature regulating" benefits.

Normally, there are always exceptions to the rules. For example, in humid climates, such as the Southeast United States, night temperatures are as humid as daytime, especially during summer. Thus, earthy building materials that absorb daytime heat release the absorbed heat into the interior space at night, even when the heat is not wanted. This creates a much warmer internal temperature, possibly leading to overheating. Although this challenge can be solved with adequate ventilation at night to carry away stored energy, it is a clear demonstration of how "one size does not fit all". It is critical that we understand our local climate and have a great understanding of passive solar design to embrace nature's gifts and the challenges that come with it. Below is a diagram of a home designed with passive solar design in mind.

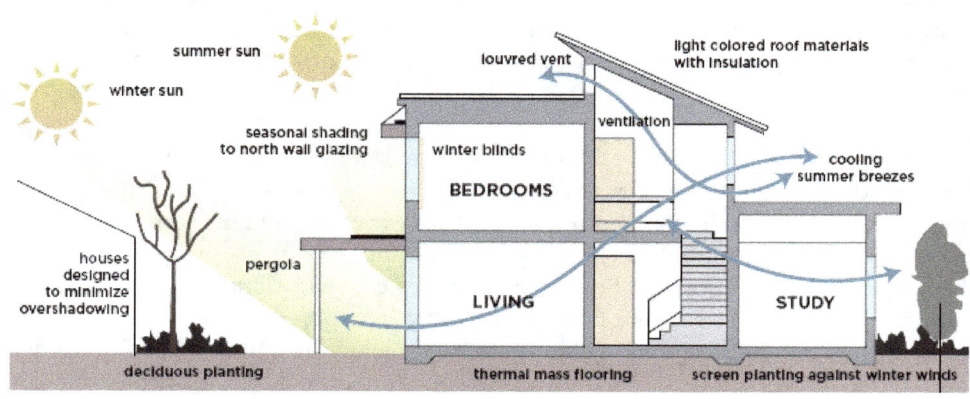

Renewable Energy   Energy Efficiency

Buildings efficiency: New techniques

Insulation: Reducing the amount of air that leaks in and out of a home is a cost-effective way to cut heating and cooling costs, improve durability, increase comfort, and create a healthier indoor environment. Proper insulation helps control airflows and can be applied either during new construction or on existing buildings. In many parts of the world, insulation made from fiber, foam, and other modern materials is readily available.

In the United States, insulation levels are specified by the R-value, which is a measure of the insulation's ability to resist heat flow. The higher the R-value, the better the thermal performance of the insulation. For example, the attic is one of the most important and most accessible places to add insulation to improve energy efficiency. Since the sun's pathway is directly overhead during summer, insulation in the attic serves as the last defense between the heat-stricken roof material and the living space below. Of course, roof material plays an important role in absorbing or transferring heat from the sun into the attic below. In most hot climate areas, rooftops made from clay tiles are very common for the same reasons discussed above. Tile rooftops absorb the sun's energy and minimize the heat transferred to the attic below. The less heat transferred to the attic, the less insulation needed. The recommended level for most attics is R-38 or about 10 to 14 inches thick, depending on insulation type and roof material.

In addition to the attic, walls on the outside perimeter need to be adequately insulated. The proper insulation means less hot or cold air will escape the building, allowing us more control of our interior temperature. The map and table below illustrate the thickness of insulation required in certain parts of the country.

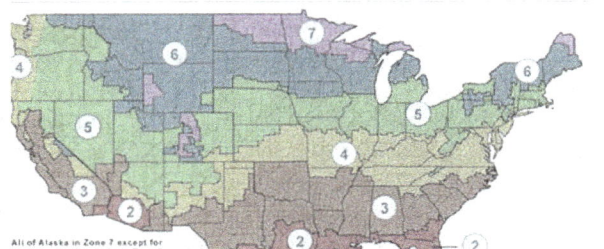

Recommended insulation levels for retrofitting existing wood-framed buildings

Source: U.S. Department of Energy

Zone	Add Insulation to Attic		Floor
	Uninsulated Attic	Existing 3–4 inches of Insulation	
1	R30 to R49	R25 to R30	R13
2	R30 to R60	R25 to R38	R13 to R19
3	R30 to R60	R25 to R38	R19 to R25
4	R38 to R60	R38	R25 to R30
5 to 8	R49 to R60	R38 to R49	R25 to R30

Air leakage: Air leakage occurs when outside air enters, and interior heated or air-conditioned air leaves the home through cracks and openings. Before applying insulation, especially in existing structures, detecting and measuring air leakage is economically smart. Detection can be accomplished visually or with measuring devices and techniques. The United States Department of Energy provides excellent recommendations for the visual detection of air leakage. Detection starts with a visual inspection of the outside of a home where there are joints or areas where two different building materials meet such as exterior corners, water faucets, where siding and chimneys meet, and areas between the foundation and the rest of the building. Any cracks in these joints are weak spots for air leaks. Inside the home, any cracks and gaps that could cause air leaks must be inspected. Those include doors and window frames, cabling outlets, gaps around pipes and wires, fireplaces, and attics.

Once air leak areas are identified, caulking and weather-stripping are two simple and effective air-sealing techniques that offer quick returns on investment. Caulk is generally used for cracks and openings between stationary house components such as around door and window frames. On the other hand, weather-stripping is used to seal components that move, such as doors and operable windows.

Following the detection of air leaks and the assessment of ventilation needs for indoor air quality, the sealing process can start. Below is a list of recommendations from the U.S. Department of Energy.

- Weather-strip doors and windows that leak air.
- Caulk and seal air leaks where plumbing, ducting, or electrical wiring comes through walls, floors, ceilings, and soffits over cabinets.
- Install foam gaskets behind the outlet and switch plates on the walls.
- Inspect dirty spots in your insulation for air leaks and mold. Seal leaks with low-expansion spray foam made for this purpose and install house flashing if needed.
- Look for dirty spots on your ceiling paint and carpet, which may indicate air leaks at interior wall/ceiling joints and wall/floor joists and caulk them.
- Cover single-pane windows with storm windows or replace them with more efficient double-pane low-emissivity windows. Even covering windows with plastic sheets can serve as short-term insulation during cold winter days. With a little extra investment, we can also benefit from technologically advanced windows and doors with higher insulation performance and glazing.

- Use foam sealant on larger gaps around windows, baseboards, and other places where air may leak out.
- Cover your kitchen exhaust fan to stop air leaks when not in use.
- Check your dryer vent to be sure it is not blocked. This will save energy and may prevent a fire.
- Replace door bottoms and thresholds with ones that have pliable sealing gaskets.
- Keep the fireplace flue damper tightly closed when not in use.
- Seal air leaks around fireplace chimneys, furnaces, and gas-fired water heater vents with fire-resistant materials such as sheet metal or sheetrock and furnace cement caulk.
- Fireplace flues are made from metal and over time, repeated heating and cooling could cause the metal to warp or break, creating a channel for air loss. To seal your flue when not in use, consider an inflatable chimney balloon. Inflatable chimney balloons fit beneath your fireplace flue when not in use. They are made from durable plastic and can be removed easily and reused hundreds of times. If you forget to remove the balloon before making a fire, the balloon will automatically deflate within seconds of coming into contact with heat.

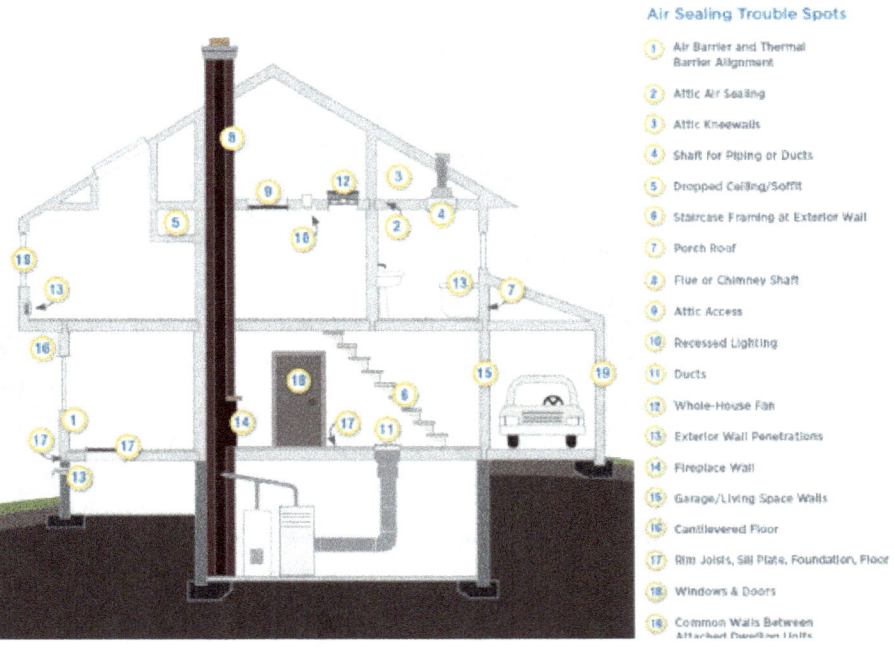

Energy Star program

Modern appliances & accessories: Modern appliances and accessories play a significant role in optimizing a passive energy concept. In the United States, the U.S. Department of Energy (DOE) classifies these accessories according to their "Energy Star". Energy Star is a certification program for manufacturers and suppliers to provide household products as energy-efficient as possible. These products help consumers save money on operating costs by reducing energy use without sacrificing performance. They also promote sustainability and reduce greenhouse emissions.

According to the DOE, Energy Star and its partners helped American families and businesses save more than 4 trillion kilowatt-hours of electricity and achieve over 3.5 billion metric tons of greenhouse gas reductions since 1992. The greenhouse gas reduction is equivalent to the annual emissions of more than 750 million cars. In 2018 alone, Energy Star helped U.S. consumers save nearly 430 billion kilowatt-hours of electricity and save $35 billion in energy costs with associated emission reductions of 330 million metric tons of greenhouse gases. Energy Star products include appliances, lighting, electronic and data equipment, heating/cooling products, windows, and doors. U.S. consumers purchased more than 300 million Energy Star-certified products in 2018. Considering a typical American household spends $2,000 a year on energy bills, Energy Star products help save up to 30% or about $575 of the energy bill.

Energy Star and real estate: DOE estimates that over 2 million Energy Star-certified new homes and apartments have been built to date, with nearly 100,000 built in 2019 alone. Today, constructing Energy Star-certified homes has become a standard process for most of America's largest homebuilders. Energy Star-certified homes and apartments are at least 10% more energy efficient. Existing buildings can also meet Energy Star certification by adding efficiency products such as insulation, windows, and doors. Commercial and industrial buildings have a lot to gain from the Energy Star program as well. On average, they waste 30% of the energy they consume. American businesses spent approximately $350 billion on energy costs in

2016 to operate commercial and industrial facilities. These businesses would save roughly $35 billion, with just a 10% improvement in energy efficiency.

U.S. Department of Energy Facts:

- The estimated annual market value of Energy Star products sales is more than $100 billion.
- On average, Energy Star-certified buildings use 35% less energy than typical buildings nationwide.
- In 2018, the Energy Star program for commercial buildings helped businesses and organizations save 190 billion kilowatt-hours of electricity, avoid $12 billion in energy costs, and achieve 140 million metric tons of greenhouse gas reductions.
- In 2018, the Energy Star program for industrial plants helped businesses save 36 billion kilowatt-hours of electricity, avoid $3 billion in energy costs, and achieve 40 million metric tons of greenhouse gas reductions.
- In 2018, the Energy Star Residential New Construction Programs helped homeowners save 3 billion kilowatt-hours of electricity, avoid $400 million in energy costs, and achieve 4 million metric tons of greenhouse gas reductions.
- 2,800 builders, developers, and manufactured housing plants are Energy Star partners, including the nation's twenty largest homebuilders. One out of every 12 single-family homes built in 2019 was Energy Star certified.
- Energy Star partners completed over 98,000 home improvement projects to increase energy efficiency and comfort in 2019, for a total of more than 873,000 to date.

LED lights: U.S. households spend over $300 billion on residential and commercial electricity bills annually, of which 15% is on lighting alone. As a result, many organizations, including the DOE, have heavily invested in lighting technology and policies to improve efficiency and reduce the cost of lighting, leading to the rapid adoption of LEDs. LED stands for light-emitting diode which is a tiny light source with a microchip. As electrical current passes through the microchip, the current illuminates the tiny light source we call LEDs. The result is a visible light. LED lighting products produce light up to 90% more efficiently than incandescent light bulbs and 80% more efficient than compact fluorescent lamps (CFL). LED lightings use heat sinks to absorb the heat produced by the LED and dissipate it into the surrounding environment. This keeps LEDs from overheating and burning out. By comparison, non-LED lights reflect the light in the desired direction, and more than half of the light may never leave the fixture leading to overheating.

Good quality LED bulbs, especially Energy Star-rated LED bulbs which are verified for quality and performance, can last more than 25 times longer than traditional light bulbs. For example, LED lights last up to 100,000 hours compared to 3,000 hours for incandescent lamps. In addition, LED lights are made from durable acrylic lenses instead of glass. This material makes LED lights much more resistant to breakage. According to the November 2015 Department of Energy report, the U.S. had around half a million LED lights installed in 2009. By 2014, that number had increased to 78 million. Today, DOE estimates at least 500 million LED recessed downlights are installed in U.S. homes and more than 20 million are sold each year.

Advancements in technology and aggressive energy policies have helped reduce the cost of LED lighting by nearly 90% since 2009. By 2030, it's estimated that LEDs will account for 75% of all lighting sales. Switching entirely to LED lights over the next two decades could save the U.S. $250 billion in energy costs, reduce electricity consumption for lighting by nearly 50%, and avoid 1,800 million metric tons of carbon emissions.

Because of their unique characteristics of high efficiency, compact size, ease of maintenance, resistance to breakage, and their directional nature, LEDs are used in a wide variety of applications. Some of the applications include recessed downlights, task lighting, traffic lights, vehicle brake lights, TVs and display cases, parking garage lighting, walkways, outdoor area lighting, refrigerated case lighting, modular lighting, and indicator lights for many electronic devices.

Outdoor displays are taken to new heights as LED lights have reduced the cost of electricity by nearly 90%. Times Square, New York, NY

Hydro Energy

Hydropower is the new and improved cousin of the old waterwheel. They both use the kinetic energy of flowing water to spin an attached element. Waterwheels convert the energy into mechanical energy by spinning an attached wheel or bucket. That mechanical energy is then used to grind grain and do other physical work. Hydropower converts the kinetic energy of flowing water to electricity by using an attached generator. This is similar to how electricity is generated from coal or natural gas power plants. Both coal and natural gas are used as fuel to boil water. The resulting steam is then used to spin a turbine. The turbine is attached to an electricity-producing generator via a shaft. Hydropower uses the energy of flowing water to spin a similar turbine and generator to produce electricity.

In theory, a dam is built on a large river that has a large drop in elevation. The dam stores lots of water behind it in a reservoir. Near the bottom of the dam wall is the water intake. Gravity causes water from the reservoir to fall through the penstock inside the dam. The turbine propeller at the end of the penstock is spun by the moving water. The propellers are attached to an electricity-producing generator via a shaft.

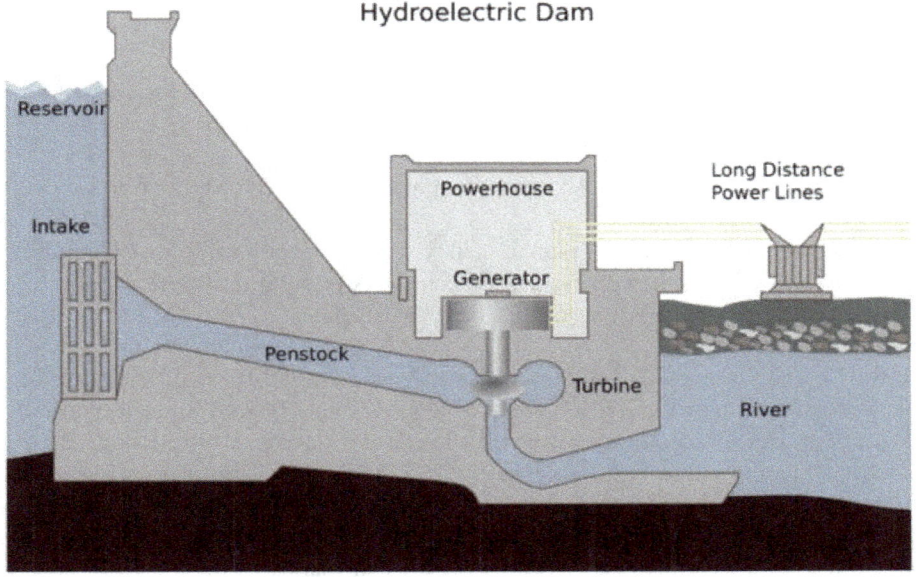

According to the U.S. Army Corps of Engineers (USACE), "A hydraulic turbine converts the energy of flowing water into mechanical energy. A hydroelectric generator converts this mechanical energy into electricity. The operation of a generator is based on the principles discovered by Faraday. He found that when a magnet is moved past a conductor, it causes electricity to flow. In a large generator, electromagnets are made by circulating direct current through loops of wire wound around stacks of magnetic steel laminations. These are called field poles and are mounted on the perimeter of the rotor. The rotor is attached to the turbine shaft and rotates at a fixed speed. When the rotor turns, it causes the field poles (the electromagnets) to move past the conductors mounted in the stator. This, in turn, causes electricity to flow and voltage to develop at the generator output terminals." The electricity is then carried to consumers via power lines. In the meantime, the water continues past the propeller through the tailrace into the river past the dam. An American icon, the Hoover Dam, built on the Colorado River bordering Nevada and Arizona, is an example of a hydroelectric dam with a large reservoir behind it.

Hydropower facilities also come in smaller dams and even dam-less, exploiting the less powerful water flows of irrigation ditches and municipal water facilities and thus, eliminating the expenses of building a dam. Hydropower facilities range from large power plants that supply many consumers with electricity to small and micro plants that individuals operate for their own energy needs or to sell power to utilities. The Department of Energy (DOE) defines hydropower sizes as follows:

> Large Hydropower: Facilities that have a capacity of more than 30 megawatts (MW).
>
> Small Hydropower: Facilities that generate 10 MW or less of power.
>
> Micro Hydropower: A small facility with a capacity of up to 100 kilowatts (KW). A small or micro-hydroelectric power system can produce enough electricity for a home, farm, ranch, or village.

In the United States, over 1,400 conventional, 40 pumped-storage, and other forms of hydropower plants operate to generate slightly over 100 GW of power. As of 2022, hydropower accounted for about 6% of total electricity output and 28% of all electricity generated via renewables in the U.S. In fact, the largest power plant of any fuel type in the United States today is the Grand Coulee hydroelectric plant in Washington State. Grand Coulee generates over 7 GW of power, twice the capacity of the second largest plant, the Palo Verde nuclear plant in Arizona.

Largest U.S. electricity generation facilities (power plants) by electricity generating capacity (2022)

Rank	Facility name	Primary fuel/energy source	State	Summer capacity (megawatts)
1	Grand Coulee	Hydroelectric	Washington	7,079
2	Palo Verde	Nuclear	Arizona	3,937
3	West County Energy Center	Natural gas	Florida	3,777
4	W.A. Parish	Natural gas, coal	Texas	3,690
5	Browns Ferry	Nuclear	Alabama	3,662
6	Bowen	Coal	Georgia	3,200
7	Gibson	Coal	Indiana	3,132
8	Monroe	Coal, petroleum	Michigan	3,080
9	Crystal River	Coal, natural gas	Florida	3,020
10	Bath County	Hydro	Virginia	3,003

Source: Form EIA-860 database, final data for 2022.

Although hydroelectric accounts for 28% of U.S. total electricity production from renewables, over half of the output is concentrated in a few states. Washington state, California, Oregon, and New York make up 60% of all hydroelectric power in the country, while the other 46 states combined for the remaining 40% of the hydroelectric power share.

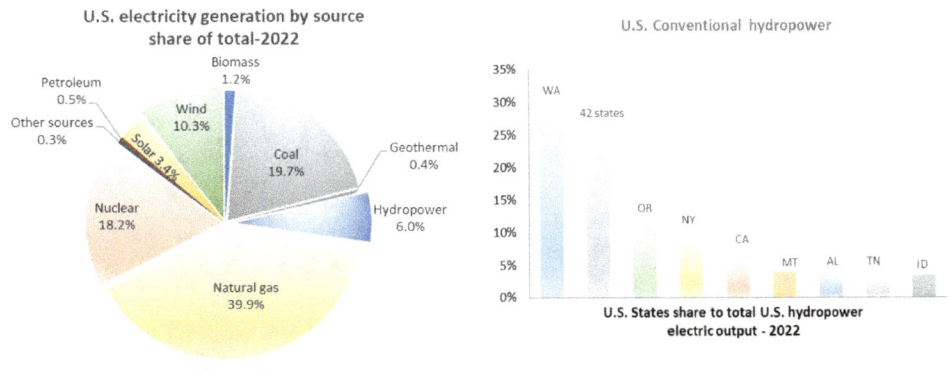

Hydroelectric power is one of the few locally sourced energy sources available. Therefore, regions blessed with an abundance of flowing water are the beneficiaries.

For example, the Northwest United States benefits from the Columbia River that flows from British Columbia, Canada, past the states of Washington and Oregon and as far south as California and as far west as Montana and Idaho. These states account for over half of the hydroelectric power generated in the United States. Some of the largest hydroelectric dams, such as Grand Coulee and Chief Joseph Dam, are in this region.

The Columbia River is the fourth-largest river in the United States by volume. The river forms in the Rocky Mountains of British Columbia, Canada. It flows northwest before flowing south into the U.S. past Washington state and Oregon for over 1200 miles (2000 km) before emptying into the Pacific Ocean. The river basin extends into additional U.S. territories in Montana, Idaho, Utah, Wyoming, and Nevada.

The steep gradients of the Columbia River and its tributaries make it ideal for hydroelectric projects. The river also has a uniform drop rate of about two feet per mile making it possible to build many dams along its 1200-mile route. In fact, there

are over 150 hydroelectric projects of all sizes on the Columbia River and its tributaries. Furthermore, the Columbia River flows through rocky canyons for much of its course. The solid rock of the canyons provides excellent footing for dams. There are 14 major dams on the mainstem river of the Columbia River. Three of these major dams are in British Columbia, Canada and eleven in the United States. Grand Coulee Dam, located in Washington state, is the largest of them all. Grand Coulee is also the largest power plant in the U.S. and the 6th largest hydroelectric plant in the world, generating over 7 GW of electricity. Meanwhile, Willamette Falls is one of the smallest power plants in the area generating 100 megawatts (MW) of electricity. Other hydroelectric plants in the area, such as Chief Joseph and John Day generate over 2 GW of power each. Overall, the hydroelectric projects on the Columbia River and its tributaries generate 36 gigawatts (GW) of hydroelectric power, enough to meet the power requirements of ten cities the size of Seattle WA. Together, they account for over 50% of all U.S. total hydroelectric generation.

The southeastern region of the United States also benefits from some of the 49 power and non-power-generating dams built on the Tennessee River. Although some of these dams are built for flood control, irrigation, and recreation only, there are 29 conventional and 1 pumped storage dams generating over 14 GW of electricity for the region.

The Tennessee river basin covers 7 states in the southeastern region of the U.S.

Despite being the most reliable and one of the lowest cost of renewable energy, hydropower has not received the kind of investments allocated to other renewable sources, such as wind and solar. As a result, electricity from hydropower has been capped at about 300 billion kWh since the early 1970s. Wind and solar powers, on the other hand, have spiked up noticeably, as illustrated in the chart below. Globally, over 1100 GW of hydroelectric power is generated accounting for 15% of all electricity output, twice as large a percentage as in the United States.

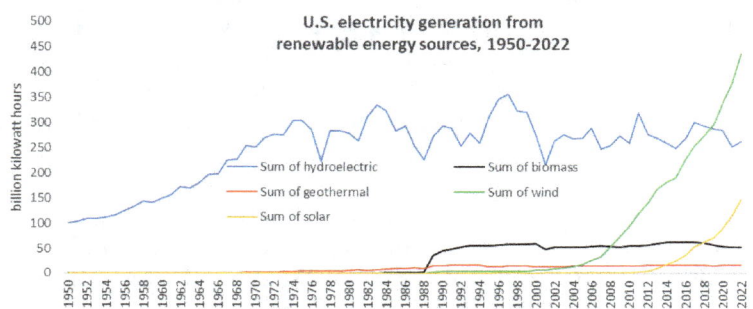

Hydropower also accounts for over 50% of all electricity generated from renewables globally. As with most renewable energy-related projects, China leads the way with nearly 30% of all global hydroelectric output.

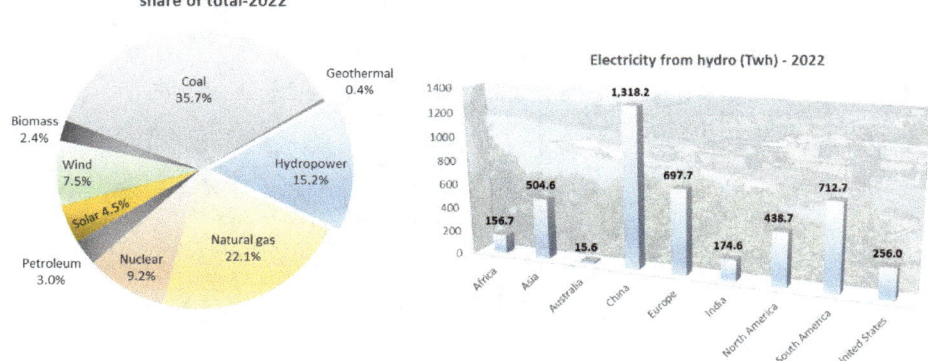

The future of hydroelectric power appears to be bright as other regions of the world continue to develop their energy sector. For example, the continent of Africa is home to two major rivers. The Nile, which is the longest river in the world, and the

Congo, the third largest (volume discharge) in the world. Unfortunately, political instability and lack of cooperation among nations have hampered any sizeable hydroelectric project in the past. Recently, some African countries are paving the way and heavily investing in their hydropower resources. Ethiopia, for example, has invested over $6 billion to build the Grand Ethiopian Renaissance Dam, capable of generating over 6 GW of electricity when completed. This would make the Ethiopian project the largest hydroelectric plant in Africa and the seventh largest in the world.

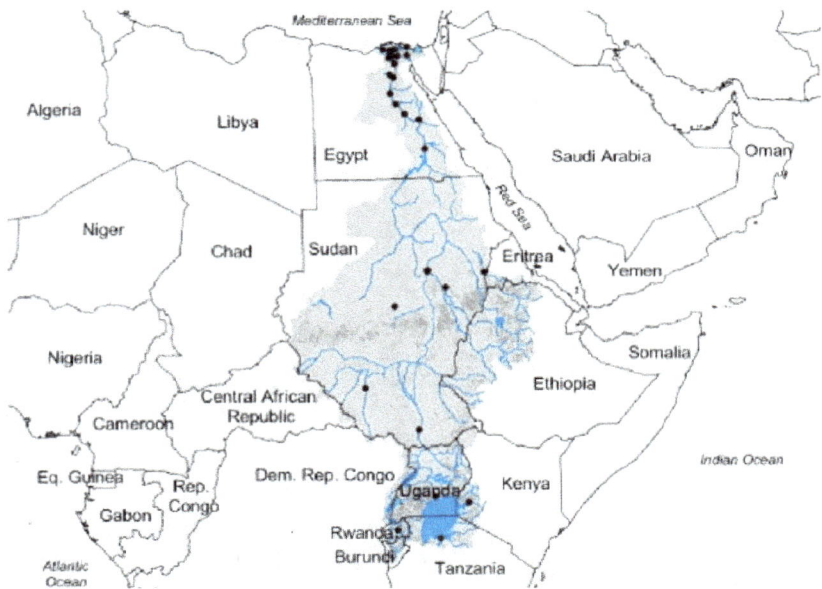

The Nile river basin which includes 11 countries remains underdeveloped with only few hydropower projects

Economics

Economically, hydropower contributes a small percentage to the overall U.S. and global employment. In 2021, the sector created slightly over 64,000 direct jobs in the U.S. and nearly 2 million jobs worldwide. These numbers are insignificant compared to the employment generated by other renewable sources, like solar and wind, which contribute over 400,000 jobs in the U.S. and over 4 million jobs worldwide annually.

Hydropower, however, has the advantage of being the most reliable and one of the cheapest sources of electricity. Obviously, significant upfront costs are unavoidable. Construction-related costs for building dams, tunnels, and other infrastructure as well as the cost of equipment, such as generators and machinery, require large initial investments. Fortunately, the long lifespan of a hydroelectric project spreads the upfront costs over time. In addition, the reliability of equipment used at hydropower facilities and the relatively low costs of maintenance, operations, and fuel make hydroelectric power very affordable. Unsurprisingly, regions blessed with an abundance of flowing water have the lowest cost of electricity anywhere in the world. For example, in the United States, Washington state and Oregon, who are beneficiaries of the Columbia River, have lower energy bills than the rest of the country. According to the U.S. Chamber of Commerce, the average cost of electricity in the U.S. is nearly 12 cents/kwh for all sectors and about 15 cents/kwh for residential. As shown in the chart below, both Washington State and Oregon residents enjoyed lower costs of electricity than the average U.S. resident for decades.

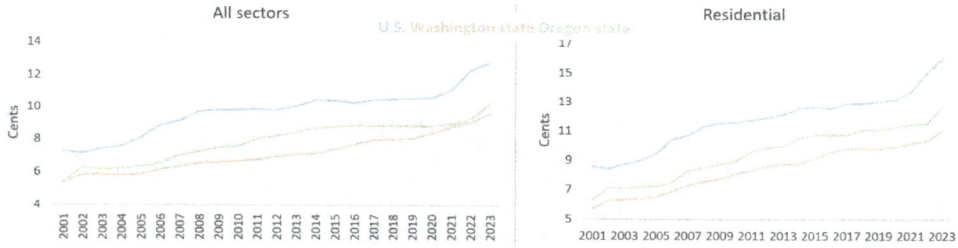

Variations

There are three types of commonly used hydropower plants in operation today: Impoundment, Diversion, and Pumped Storage.

Impoundment: The most common type of hydroelectric power plant is an impoundment facility. An impoundment facility is typically a large hydropower system that uses a dam to store river water in a reservoir. Water released from the reservoir flows through a turbine and spins the attached generator to produce electricity. In the end, the water is either allowed to continue down the river or back to the reservoir to maintain a constant reservoir level.

For impoundment facilities, electricity generation capacity is heavily dependent on the amount of flowing water and the droppage height of the water. For example, a large amount of flowing water spins the turbines with considerable force thus generating more electricity. At the same time, the deeper the water drops, the greater the impact force will be on the turbines, leading to more electricity generation. Thus, larger reservoirs and taller dams have a greater capacity for generating electricity. One of the earliest projects to incorporate both a large reservoir and a tall dam height is the Hoover Dam. At 726 feet (221 meters) tall, Hoover Dam is one of the tallest hydroelectric plants constructed. The dam is built on the Colorado River bordering the states of Arizona and Nevada. At its completion in 1936, Hoover Dam was the largest hydropower facility in the world. Currently, the facility hosts a total of 17 Francis turbines generating over 2 GW of electricity to millions of homes and businesses in Arizona, Nevada, and southern California.

Water is fed to the turbines through the four water intake towers. Two towers on the Arizona side and two towers on the Nevada side. The towers are 395 feet tall each and are made of reinforced concrete and steel. The intakes divert large volumes of water to the turbines below, through 30-ft diameter concrete lined pipes/penstocks.

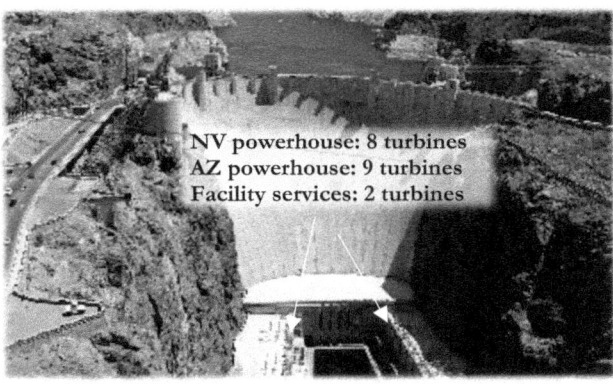

NV powerhouse: 8 turbines
AZ powerhouse: 9 turbines
Facility services: 2 turbines

NV side: 2 intake towers
AZ side: 2 intake towers

Renewable Energy   Hydro Energy

The construction of Hoover Dam was possible because of one of the largest man-made reservoirs in the United States, Lake Mead. Lake Mead is formed to back up into the Colorado River and serves as a reservoir for Hoover Dam. The lake was formed by the Hoover Dam project on September 30, 1935. In addition to serving as a reservoir for Hoover Dam, the lake serves water to nearly 20 million people and large areas of farmland in Arizona, California, Nevada, and parts of Mexico. Recreationally, it is a Mecca for outdoor enthusiasts who love to swim, boat, hike, cycle, camp, and fish.

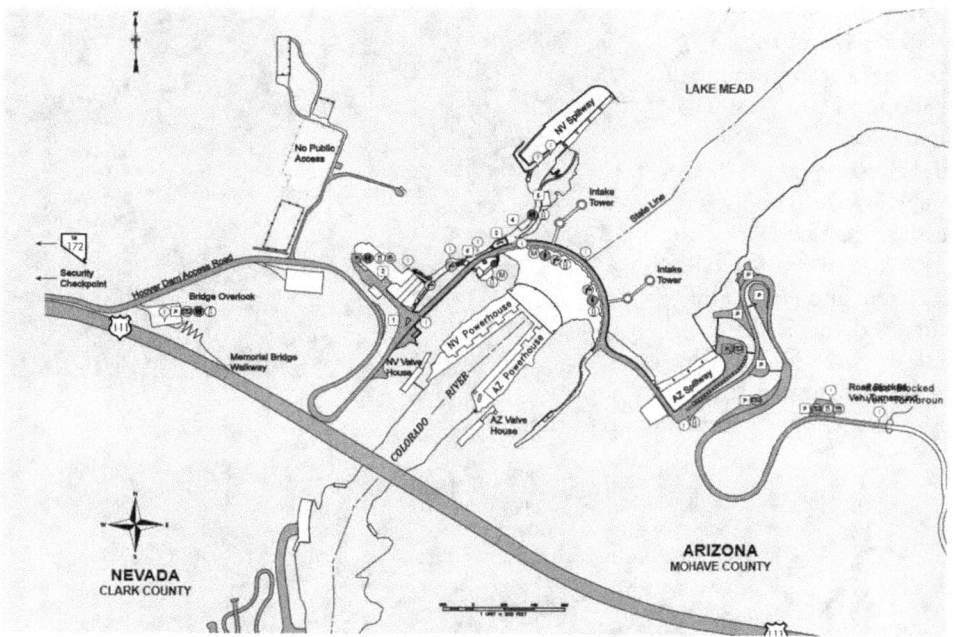

Impoundment hydroelectric facilities lead to large water reservoirs
Lake Mead: An artificial lake created as a reservoir for Hoover Dam

- Formed by the damming of the Colorado River,
- Lake Mead extends 115 miles (185 km) upstream,
- from 1 to 10 miles (1.6 to 16 km) wide,
- has a capacity of 31,047,000 acre-feet (38,296,200,000 cubic meters),
- has 550 miles (885 km) of shoreline,
- has a surface area of 229 square miles (593 square km).

Hoover Dam intake towers and spillways Diagram

Source: National park services – U.S. Department of the interior

Spillways

Concrete-lined open channels about 650 feet long, 150 feet wide, and 170 feet deep on each canyon wall.

Maximum water velocity in the spillway channels is about 175 feet per second or 120 miles per hour.

Each spillway can discharge 200,000 cfs

If the spillways were operated at full capacity, the energy of the falling water would be about the same as the flow over Niagara Falls. The drop from the top of the raised spillway gates to the river level would be approximately three times as great.

Another example of an impoundment dam is the Bonneville Dam near Portland, Oregon. According to the US Army of Corps of Engineers, Bonneville Dam is 197 feet tall, making it one of the shortest in the United States. By comparison, the Hoover Dam is 726 feet tall. Bonneville Dam was completed in the late 1930s to provide hydropower to neighboring cities in the Pacific Northwest. The initial powerhouse was built with a capacity of up to ten generators producing nearly 0.5 GW of power. Later, in 1981, an additional powerhouse was built for an additional capacity of eight generators. Today, the facility has 20 turbines and generates nearly 1.2 GW of power, enough electricity to power about a million homes. Today, the project serves additional benefits, such as river navigation, flood risk management, irrigation, fish and wildlife habitat, and recreation.

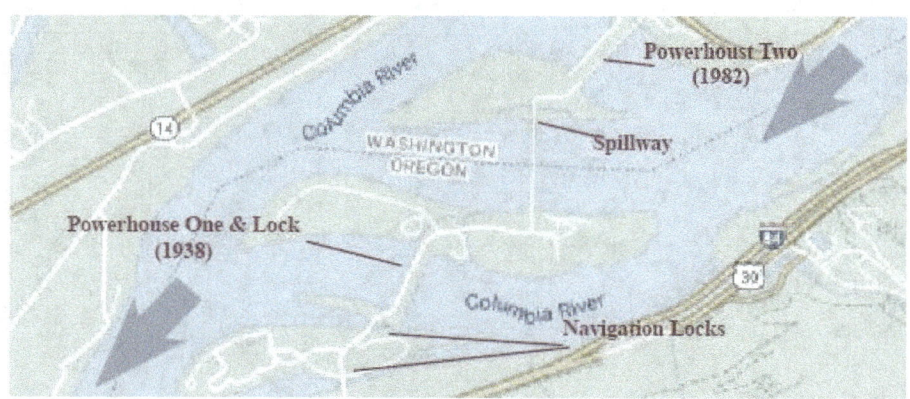

The Bonneville project serves the additional benefits of river navigation, flood control, irrigation, and recreation.

Bonneville is one of the shortest dams in the U.S. generates 1.2GW of power with 20 turbines

Since the early 1930s, an additional 13 dams have been built on the Columbia River. Overall, there are 60 dams in the watershed, including 14 on the Columbia, 20 on the Snake, 7 on the Kootenay, 7 on the Pend Oreille/Clark, 2 on the Flathead, 8 on the Yakima, and 2 on the Owyhee, combining for a total of over 36 GW of electric power. Grand Coulee Dam, at only 550 ft high, is the largest producer of hydroelectric power in the river basin and is the largest hydropower facility in the United States, generating over 7 GW with 33 turbines. Willamette Falls Dam, completed in 1888, is one of the smallest in the river basin at a capacity of 15 MW at 20 ft high.

Hydroelectric power on the Columbia River accounts for over 50% of electricity generation for both Oregon and Washington state. In fact, Oregon and Washington account for over a third of all US hydropower. In addition, the state of Oregon generates 11% (3.4 GW) of electricity from wind power from more than 1,900 turbines. Today, Oregon produces over 60% of its electricity from renewable sources. With the addition of more wind farms, solar power, and other renewables, Oregon is pushing for 100% renewable by 2045. As for Washington state, nearly 67% of its electricity comes from hydroelectric power from the Columbia River. The state's prized joule is the Grand Coulee hydroelectric dam, which is the largest hydroelectric plant in the U.S. with a capacity of 7 GW, enough to power 4.2 million households annually.

Pumped storage: The southeast region of the United States also comprises multiple hydroelectric projects along the Tennessee River valley. The region has over 29 hydropower dams generating electricity for neighboring states. One of the largest power-producing plants in the area is the "Raccoon Mountain Pumped Storage plant". Pumped storage hydropower plants store energy using a system of two interconnected reservoirs with one at a higher elevation than the other. The plant works similarly to an impoundment facility. During periods of high electricity demand, the water at the upper reservoir is released to the lower reservoir

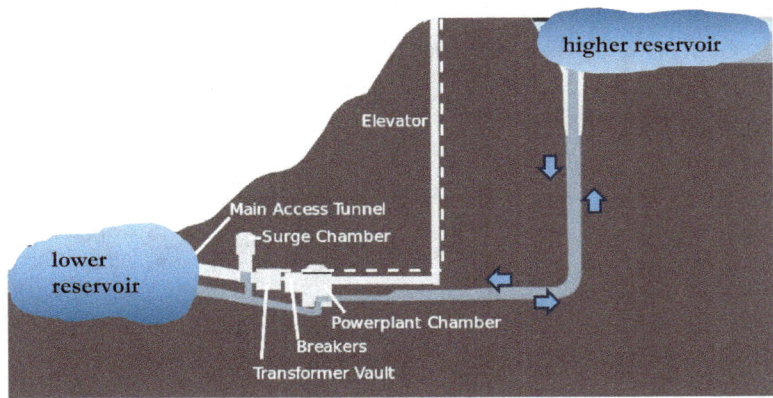

generating electricity. During excess energy time, the water is directed back to the upper reservoir and stored for future use. Pumping the water to the upper reservoir is managed using excess electricity from the grid during low electricity demand. The process is accomplished using a reversible Francis Turbine/Pump, which is specifically designed for this purpose.

According to Tennessee Valley Authority (TVA), which owns the facility, Raccoon Mountain Pumped Storage works like a large storage battery. During periods of low demand, water is pumped from Nickajack Reservoir at the base of the mountain to the reservoir built at the top. The reservoir contains approximately 107 billion gallons of water covering 528 acres of water surface and takes 28 hours to fill. When demand is high, water is released via a tunnel drilled through the center of the mountain to drive generators in the mountain's underground power plant. The dam at Raccoon Mountain's upper reservoir is 230 feet high and 8,500 feet long.

The facility has four generators producing 413 MW each, for a total of 1.6 GW net capacity. Net capacity is the amount of power the plant produces on an average day, minus the electricity used by the plant itself. The facility was completed in 1978 after 8 years of construction. Today, the plant is used most days and serves as an important element for peak power generation and grid balancing in the TVA system. Globally, Raccoon Mountain plant is considered one of the largest power-producing plants. Additionally, the facility provides great recreational values, such as hiking and mountain biking.

Francis turbine/Pump used at Raccoon Pumped Storage plant.

Pump Capacity: 519,000 Hp at 1000ft head.

Turbine Capacity: 525,000 HP at 1020ft head.

The low energy density of a pumped storage system requires either large flows and/or large differences in height between reservoirs. In some places, this occurs naturally, while in other cases, one or both bodies of water are man-made. The reservoirs can also be open-loop or closed-loop systems. An open loop system involves a lower water system that is connected to a naturally flowing river; thus, loss of water capacity is not an issue. In fact, if the upper reservoir collects significant rainfall or is fed by a river, then the plant may be a net energy producer in the manner of a traditional hydroelectric plant. Closed systems, on the other hand, involve two water storage facilities that are closed off. Thus, decreasing water

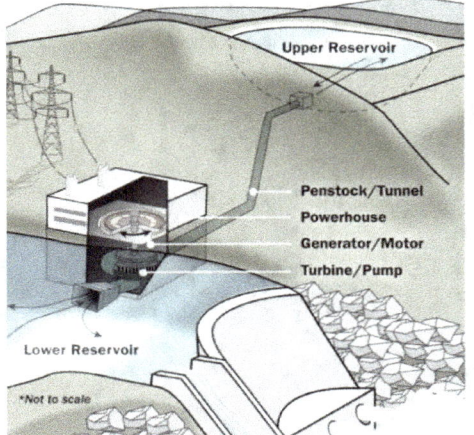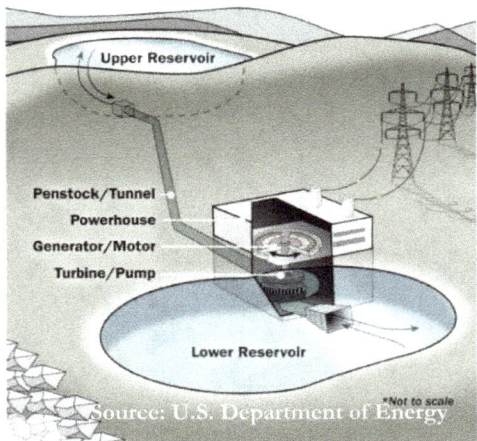

capacity could be a concern.

Because a large amount of energy is required to pump water uphill, pumped storage reservoirs are relatively small. Their small size gives them the advantage of starting up quickly. Their small sizes also make them very efficient and inexpensive. The U.S. Department of Energy Global Energy Storage Database showed a total stored energy of over 181 GW globally in 2020. China leads with 32 GW of total electricity generating capacity while Japan and the U.S. are close 2nd and 3rd at a total

Rank	Facility	Location	Total Capacity (MW)	Operation Start Date
1	Bath County	USA	3003	1985
2	Huizhou	China	2440	2011
3	Guangdong	China	2400	1994, 2000
4	Okutataragi	Japan	1932	1974
5	Ludington	USA	1872	1973
6	Tianhuangping	China	1836	2004
7	Tumut-3	Australia	1800	1959
8	Grand'Maison Dam	France	1800	1985
9	La Muela II	Spain	1772	2013
10	Dinorwig	UK	1728	1984
11	Raccoon Mountain	USA	1652	1978

generating capacity of 28 GW and 22 GW, respectively. Below is a list of the largest pumped storage facilities in the world

Diversion (run-of-the-river): A diversion facility, also known as a run-of-river facility (ROR), channels a portion of the river water through a canal or penstock to an electricity-generating powerhouse before the water rejoins the main river further downstream. It can be used with a dam or dam-less system. Thus, generating electricity in a ROR system requires little water storage, known as bondage, or no water storage at all. A plant without bondage is subject to seasonal river flows. Thus, the plant will operate as an intermittent energy source.

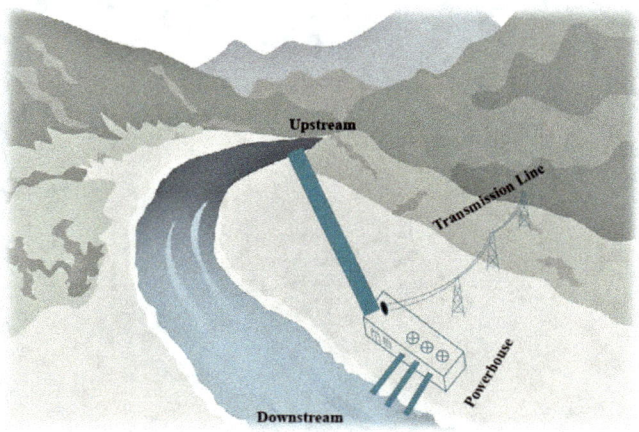

The best sites for ROR projects are areas where there is strong year-round water flow originating either from rainfall or snowpack melting and a large gravitational drop or hydrostatic head to enhance the water's energy. A greater drop in elevation means more gravitational force acts upon the water, increasing its kinetic energy.

P (power produced) = Q (water flowmeters/second) • H (hydrostatic head of water-meters) x 7.83

The disadvantage of ROR is that when the river's water levels are depleted because of drought or water extraction, the power output is reduced or becomes entirely unavailable. Basically, Q (water flow) suffers. The lack of a reservoir also puts an upper limit on the size of the run of river plants. Thus, they are only feasible on rivers with large year-round flow rates. For example, the Blue Nile which originates in the highlands of Ethiopia reaches its maximum capacity during the rainy season between June and September. On the other hand, the White Nile which originates further south in the Great Lakes region has a lower volume, but consistent flow rate throughout the year. Therefore the White Nile offers a better ROR opportunity than the Blue Nile.

Countries that have many large hydroelectric power projects are also leaders in ROR hydropower projects. For example, China, Brazil, and Canada possess the largest hydropower-generating capacity. They are also leaders in ROR power systems. In fact, more than half of the world's ROR capacity is in China. Mountainous countries with enormous hydropower resources are also investing in ROR power to top up their power generation portfolios. Nepal, Norway, Switzerland, and Austria are notable examples. In the United States, the mountainous region of the Pacific Northwest offers opportunities for ROR system as well. Both the largest (Chief Joseph Dam) and the oldest (Willamette Falls) ROR systems are located in this region on the Columbia River.

Chief Joseph Dam
Largest ROR system in the U.S.

Chief Joseph Dam is a concrete gravity run-of-the-river facility on the Columbia River, near Bridgeport, Washington. Since there is no reservoir, water flows to Chief Joseph Dam from another upstream power plant, Grand Coulee Dam. This is an example of ROR facility installed further downstream from another hydropower facility but operates independently to generate additional power. At a capacity of 2.6 GW, Chief Joseph Dam is one of the largest hydroelectric power producers in the United States. The single powerhouse is over a third of a mile long and holds 27 main unit penstocks to deliver water to each of the 27 house-sized Francis turbines. These 27 turbines/generators produce 2.6 GW of electricity, enough power to supply the whole Seattle metropolitan area. In addition, the facility has two small penstocks that supply water for two station service generators used to supply the power to operate the facility.

Willamette Falls Dam
Oldest ROR system in the U.S.

Located in Willamette Falls, Oregon City, Oregon, Willamette Falls Dam is one of the oldest and the smallest ROR facilities in the U.S. Even though the dam has a natural waterfall of only 40 feet, it has been harvested to provide hydroelectric power since 1888. Historically, the falls served power to lumber a mill, a flour mill, a woolen mill, and a paper mill since the 1840s. In 1889, Station A (later decommissioned and replaced) in Oregon City delivered power 14 miles to Portland, Oregon making it the longest-distance AC power delivery at the time. Willamette Falls consists of a 600-foot spillway section and a 2,300-foot dam topped with flashboards. The water intake for the turbines is located at the base of the powerhouse. The powerhouse (T.W. Sullivan, TWS) contains 13 turbines/generators with a total generating capacity of 16 MW. Water diverted through the powerhouse rejoins the main river immediately below the Falls. Willamette Falls operates in a run-of-the-river mode and does not provide usable water storage or flood control. However, at only 40 feet drop, this plant is an excellent model for harvesting power regardless of volume or drop height.

The are numerous advantages to ROR systems, including:

- The absence of water storage or reservoir makes these facilities very inexpensive.
- ROR systems can be installed at existing dams, as an independent generating facility, or in private systems to power small communities.

- Sometimes, plants are constructed in conjunction with river and lake water-level control and irrigation systems, as well as water flows from municipality facilities.
- The lack of a major reservoir reduces the environmental footprint of ROR plants. In large hydro projects, the creation of a reservoir affects local communities as well as plant and animal life. Standing water in a reservoir can hurt overall water quality as well. With ROR systems, these impacts are not wholly avoided, but they are minimized to what is often considered a tolerable degree.

Below is a list of the largest hydroelectric power capacity plants in the United States

Rank	Facility	Capacity (MW)	State	Operating Year	Type
1	Grand Coulee	6,809	Washington	1942	Conventional
2	Bath County	3,003	Virginia	1985	Pumped Storage
3	Robert Moses Niagara	2,675	New York	1961	Pumped Storage
4	Chief Joseph	2,614	Washington	1979	Run-of-River
5	John Day	2,485	Washington/Oregon	1971	Run-of-River
6	Ludington	2,172	Michigan	1973	Pumped Storage
7	Hoover	2,080	Arizona/Nevada	1936	Conventional
8	The Dalles	1,813	Washington/Oregon	1957	Run-of-River
9	Raccoon Mountain	1,616	Tennessee	1978	Pumped Storage
10	Castaic	1,500	California	1973	Pumped Storage

The nuts and bolts of hydropower

As previously discussed, a major component of the hydropower system is the turbine. Hydroelectric turbines come in a variety of shapes and vary considerably in size. In addition, there are some criteria to consider during turbines selection, including:

- how deep the turbine must be
- turbine efficiency
- manufacturing material and cost
- site parameters that will determine the most suitable turbine

There are two main types of hydro turbines on the market, Impulse and Reaction. The preferred turbine is considered according to the height of standing water (known as head) and the flow/volume of water.

Impulse turbine: used for high head (>1000ft) and low flow systems. Impulse turbine generally uses the velocity of the water to move the runner and discharges to atmospheric pressure. The water jet hits each bucket and pushes on the turbine's curved blades, which changes the direction of the flow. The resulting change in momentum (impulse) causes a force on the turbine blades. There is no suction on the downside of the turbine, and the water flows out the bottom of the turbine housing after hitting the runner. Examples of Impulse turbines include:

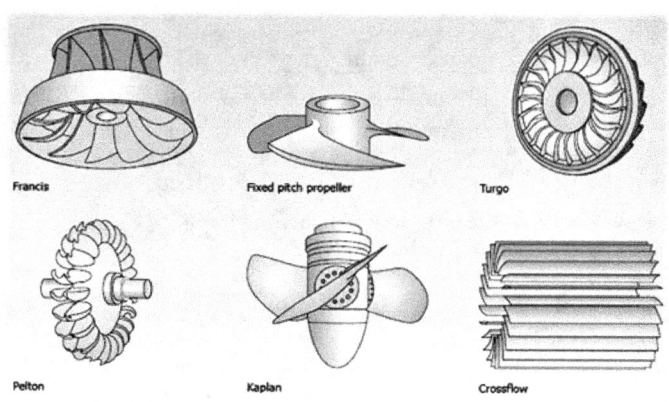

Turbines come in a variety of shapes and sizes.

Pelton Wheel: A Pelton wheel has one or more free jets discharging water into an aerated space and impinging on the buckets of a runner. Some of these turbines have a runner diameter of up to 13 ft (4 meters) for heads of 655–5900 ft (200-1800 meters). They are capable of an output of up to 350 MW.

Turgo Wheel: is a variation of the Pelton and is a cast wheel whose shape generally resembles a fan blade that is closed on the outer edges. The water stream is applied on one side, goes across the blades, and exits on the other side.

Cross-Flow: A cross-flow turbine is drum-shaped and uses an elongated, rectangular-section nozzle directed against curved vanes on a cylindrically shaped runner. The cross-flow turbine allows the water to flow through the blades twice. The first pass is when the water flows from the outside of the blades to the inside; the second pass is from the inside back out.

Water Wheel: A water wheel is a machine for converting the energy of flowing or falling water into useful forms of power, often in a watermill. A water wheel consists of a wheel (usually constructed from wood or metal), with several blades or buckets arranged on the outside rim forming the driving car. Water wheels were still in commercial use well into the 20th century but are no longer in common use. Water wheels were normally used for milling flour in gristmills, grinding wood into pulp for papermaking, hammering wrought iron, machining, ore crushing and pounding fiber for use in the manufacture of cloth.

Water wheel: West Virginia

Reaction turbine: is the second type of turbine used for low head (<100ft) and/or for medium head (100-1000ft) with high flow systems. Reaction turbines are acted on by water, which changes pressure as it moves through the turbine and gives up its energy. They must be encased to contain the water pressure (or suction), or they must be fully submerged in the water flow. In this system, the runner is placed directly in the water stream flowing over the blades rather than striking each individually. Examples of reaction turbines include:

Propeller: A propeller turbine generally has a runner with three to six blades in which the water contacts all the blades constantly. The major components include the runner, scroll case, wicket gates and a draft tube.

Bulb Turbine: The turbine and generator are sealed units placed directly in the water stream. The term "Bulb" describes the shape of the upstream watertight casing, which contains a generator located on a horizontal axis.

Straflo: The generator is attached directly to the perimeter of the turbine.

Tube Turbine: The penstock bends just before or after the runner, allowing a straight-line connection to the generator.

Francis: A Francis turbine has a runner with fixed buckets. Water is introduced just above the runner and all around it and then falls through, causing it to spin. This is by far the most common type in present-day medium or large-scale plants with heads as low as 6ft (2m) or as high as 656 ft (200m). In fact, 60% of the global hydropower capacity uses Francis turbines. Modern Francis turbines have outputs of up to 800 MW with a runner diameter of up to 33ft (10m) and weighing over 400 tons. They can also achieve efficiencies as high as 95%. Francis turbines can sustain the high mechanical stress resulting from high heads. They are also mounted with a vertical shaft to isolate water from the generator.

Francis turbine: Chief Joseph Dam

Kaplan: Both the blades and the wicket gates are adjustable, allowing for a broader range of operation. With a double regulation system, Kaplan turbines provide high efficiency over a broad range of configurations. The vertical configuration of the Kaplan turbine allows for larger runner diameters (above 33ft/10 m) and increased unit power, as compared to Bulb Turbines. Some modern Kaplan turbines are also engineered with a "fish-friendly" structure to improve the survival rate of migrating species. They are also equipped with water-lubricated bearings and water-filled hubs to prevent water pollution.

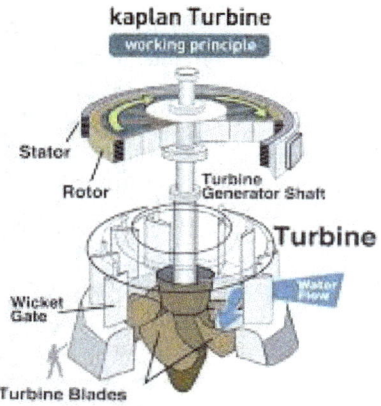

The Kaplan turbine was an evolution of the Francis turbine. Its invention allowed efficient power production in low-head applications, which was not possible with Francis turbines. The head ranges from 33 to 230ft (10 to 70 meters) and the output ranges from 5 to 200 MW. Runner diameters are between 6 and 36ft (2 and 11 meters). Turbines rotate at a constant rate, which varies from facility to facility.

Kinetic: Kinetic energy turbines are known as free-flow turbines. They use the kinetic energy of flowing water as in rivers or streams rather than the potential energy from the head. The system is ideal for application in rivers, man-made channels, tidal waters, or ocean currents. No damming or water diversion is required. The system uses the natural pathway of the water source.

Deriaz/Pump turbine: Designed for pumped storage hydroelectric. They can reverse flow and operate as a pump to fill a high reservoir during off-peak electrical hours, and then revert to a water turbine for power generation during peak electrical demand. They range from 30 MW to 400 MW per unit with heads up to 1,000m. Fast startup times of just 90 seconds for up to 400 MW allow for an increased number of daily starts and stops, adding flexibility and availability.

Lastly, turbine blades need to have high corrosion resistance and strength because they are constantly exposed to water and extreme water pressure. Thus, most blades are made of martensitic stainless steels, which have high strength compared to austenitic stainless steels. In addition, a small percentage of chromium is added to the steel alloy to increase corrosion resistance and expand the lifespan of the blades. Martensitic stainless-steel alloys also have low density and thinner sections. These qualities allow the blades to rotate more easily and lead to overall turbine efficiency. Finally, the turbines also need a higher weld quality for easier repair.

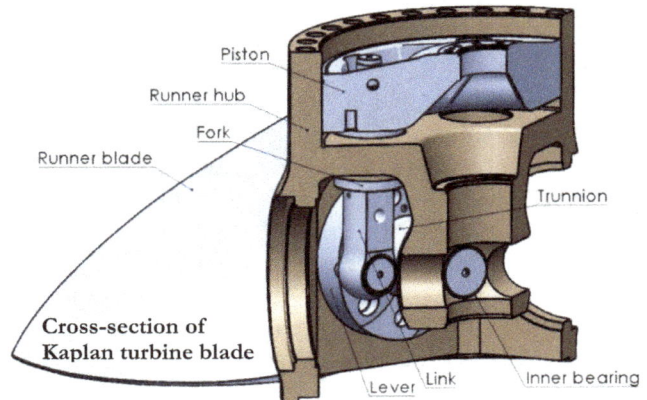

Hydropower benefits

Hydropower has been recognized and harnessed since the late 1800s. It is a clean and cost-effective form of energy. Additionally, since hydropower is a domestic source of energy, it allows regions to produce energy independently. Hydropower is more widely used than most people realize. In the U.S., all but two states (Delaware and Mississippi) generate electricity from hydropower. According to the U.S. Army of Corp Engineers, hydropower has the following additional benefits:

> Hydropower is more efficient than any other form of electrical generation. It can convert 90 percent of available energy into electricity. The best fossil fuel plant is only about 50 percent efficient.
>
> Hydropower is a low-cost alternative. On average, hydropower production costs one-third that of nuclear or fossil fuel production.
>
> Hydropower can easily respond to power needs. Hydropower dams can be turned on and off quickly. Other forms of electricity production, such as coal power, require a great deal of time to start or stop producing electricity.
>
> Hydropower is a clean, reusable source of electricity. It produces no emissions, and its fuel (water) can be used at each downstream dam.
>
> Hydropower is domestic and the supply of water is continually replenished through rain and snowmelt. They are immune from foreign supply interruption.

Dams and reservoirs designed for hydropower also serve other purposes, such as flood control, irrigation, drinking water supply, fishing, and various recreational opportunities. In fact, dams and reservoirs were not originally built for hydropower purposes. For example, there are over 80,000 dams/reservoirs in the United States but only 2500 or 3% are used for producing electricity. The other non-hydropower dams represent untapped potential for clean, renewable energy. Similar opportunities for hydropower exist all over the world. Many countries are accustomed to building dams/reservoirs for irrigation and other purposes. Those projects remain unexploited for hydropower as well.

For example, the country of Eritrea on the African continent, has heavily invested in man-made lakes to develop its agricultural economy and achieve national food security. Over the last three decades, the country has built over 700 micro/small, medium, and large lakes. According to the Ministry of Agriculture, the investment has helped expand the nation's arable land and has helped achieve food security. In addition, the reservoirs contribute to the nation's drinking water supply while providing an opportunity for the local fish economy.

In the near future, some of these dams and reservoirs can be used to generate electricity for local demand or to add to the national electricity grid system. Their mountainous location and variant sizes make them ideal for pumped storage and ROR systems. Whether it is in the United States or Eritrea, a large percentage of dams and reservoirs remain untapped potential. The availability of cost-effective systems and efficient equipment should contribute to turning some of these dams into electricity-producing assets for our energy-hungry planet.

Some of the larger man-made lakes in Eritrea suitable for Pumped Storage or ROR systems

Wind Energy

Wind energy has been utilized for thousands of years. Although the mechanism of energy production has changed, the concept has been consistent for generations. Wind Energy is one of the oldest technology known to man. As previously mentioned, Egyptians used wind power to propel boats along the Nile River as far back as 5000 BC. Windmills with woven-reed blades were used for grinding grain in Persia, the Middle East, and China centuries ago. Wind pumps have been used to pump water since the 9th century in Afghanistan, Iran, and Pakistan, and later spread to China and India. By the 11th century, wind pumps and windmills were used extensively for food production worldwide, including in Europe, where merchants and crusaders introduced the technology. In Europe, particularly in the Netherlands and in the East Anglia area of Great Britain, windmills were used to drain land for agriculture.

A few centuries later, European immigrants brought wind energy technology to the Western Hemisphere, where windmills were used to grind grain, pump water, and cut wood at sawmills. Ranchers and farmers installed thousands of wind pumps as they settled the western United States. With the increased electrification of rural communities in the 1930s, wind pumps and windmills were less utilized. The era of electrification led to modern wind energy technology where the wind power needed to be converted to electricity.

Wind energy boomed in the wake of oil shortages and climate change concerns in the 1970s. In response to the oil embargo of 1970 and the Iranian revolution of 1979, the United States, along with other nations, started investing in the development of alternative energy sources. The U.S. federal government provided research and development funding to reduce the cost of wind turbines and offered tax and investment incentives for wind power projects. By the early 1980s, thousands of wind turbines were installed in California, largely supported by federal and state policies that encouraged renewable energy sources.

A few years later, the federal government established incentives to use renewable energy sources in response to renewed concerns over climate change and fluctuating fuel prices. Congress passed the Production Tax Credit (PTC) in 1992 and the Investment Tax Credit (ITC) in 2006 to incentivize the production and distribution of electricity from renewable sources. In addition, state governments enacted new requirements for electricity generation from renewable sources. Soon, electric power marketers and utilities began to offer electricity generated from wind and other renewable energy sources, called green power, to their customers. The tax credits for wind and solar anchored by advances in technology contributed to a remarkable drop in the cost of renewables projects. In the past decade alone, wind power costs have declined by nearly 70%. As a result, U.S. total annual Electricity generation from wind power increased from 6 billion kilowatt-hours (kWh) in 2000 to over 400 billion kWh in 2022, contributing to nearly 10.3% of total U.S. utility-scale electricity generation and 29% of all renewable sources consumed in the United States.

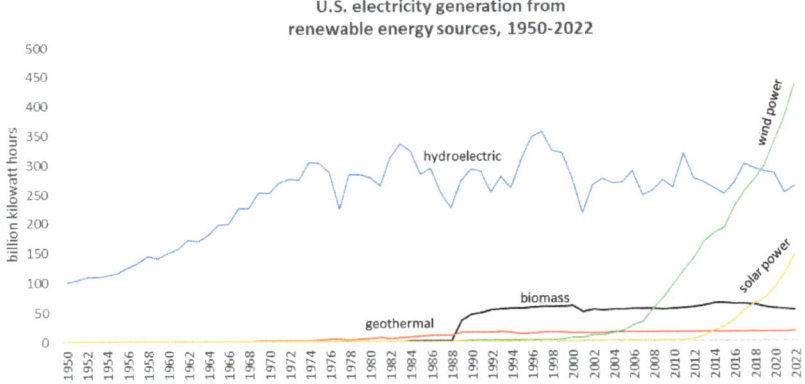

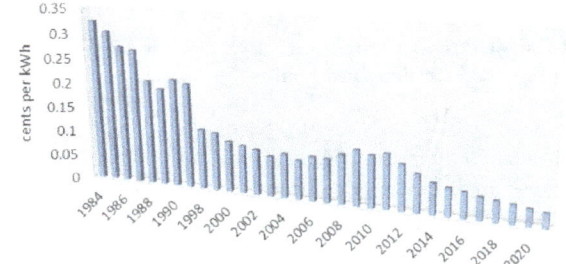

Cost of electricity from wind has decreased from 30 cents per kWh in the 1980s to below 5 cents per kWh today. Wind power is now the 2nd cheapest source of electricity in the U.S.

Most importantly, wind power projects were very dispersed across the U.S. with over 42 states involved in some capacity. The five states with the most electricity generated from wind in 2019 were Texas, Oklahoma, Iowa, Kansas, and California. These states combined for nearly 60% of total U.S. wind electricity generation in 2019. Table 3 below illustrates each state's wind power capacity as of 2019.

Table 3: Top 5 states for wind power capacity - 2019

Rank	State	Electricity generation capacity (MW)	Highlights
1	Texas	24,899	Top Wind Farm: Roscoe wind farm with 634 turbines. Enough to power nearly 6 million homes
2	Iowa	8,422	4637 turbines contribute to 37% of state's electricity
3	Oklahoma	8,072	Generates 31.9% of its electricity from wind power
4	California	5,885	Alta wind energy center: Largest wind farm in the U.S. and 2nd largest in the world. Top Solar electricity producer in the U.S. Commited to 100% renewable energy by 2045
5	Kansas	5,653	3000 turbines. Commited to 20% electricity generation from wind power by 2020

Globally, wind power has also increased substantially in recent years. In 1990, 16 countries generated a total of about 3.6 billion kWh of wind electricity. In 2000, 49 countries generated about 31 billion kWh, and in 2017, 129 countries generated about 1,129 billion kWh of wind electricity

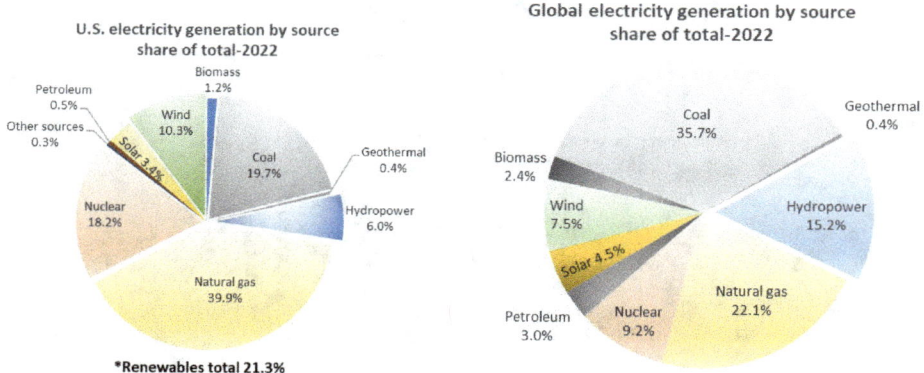

Wind power accounts for 10.3% of all U.S. electricity production. Globally, wind power accounts for 7.5% of global electricity production

In addition to contributing to cheaper and cleaner electricity, wind energy has also brought in direct investments and created new job opportunities for the U.S. economy. According to the American Wind Energy Association (AWEA), wind energy, supported by U.S. tax incentives and credits of the past decade, has created:

- 114,000 jobs and generated more than $143 billion in private investment across all 50 states.
- Contributed more than $760 million in local and state taxes.
- Resulted in nearly $290 million in lease payments to farmers annually.

Wind Electric Power Generation jobs - Employment by Industry Sector

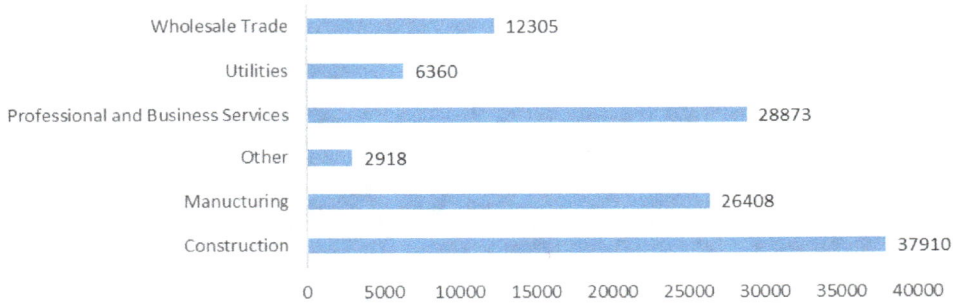

Wind energy is also creating new opportunities in factory towns across America. Over 530 factories across 43 states build wind energy-related parts. The above chart illustrates the total employment numbers in the electricity generation sector. The wind energy sector employs the second most renewable energy at 114K jobs in the U.S. and employs over 11 million people globally. Most importantly, since most of the jobs are in the construction and utility sectors, the jobs will remain locally.

Understanding wind energy

Wind energy is the process of creating electricity using the power of the wind and wind turbines. The process crosses three energy conversions: The kinetic energy of the wind spins the blades. The blades spin the attached internal shaft creating mechanical energy. Finally, the internal shaft spins an attached generator, which produces electrical energy or electricity. Electrical wires and cables then transport the generated electricity to the grid for distribution.

Wind turbines start generating electricity when wind speeds reach 6 to 9 miles per hour. This starting wind speed is known as the cut-in speed. As wind speeds increase, so does electricity production. However, a wind speed of about 55 mph and above is considered too dangerous, and the turbines are shut down to prevent equipment failure. Modern wind turbines have an average life of over 25 years and can generate usable amounts of electricity over 90 percent of the time.

The process of converting wind to electricity (illustrated below) requires massive initial investment. The components required to generate electricity are numerous. The tower, the blade, and the nacelle are the biggest factors in determining the amount of electricity a wind turbine can generate. Bigger blades on a taller tower can capture more wind to run a bigger generator. The largest wind turbines in operation today have electricity-generating capacities of up to 10 megawatts (MW).

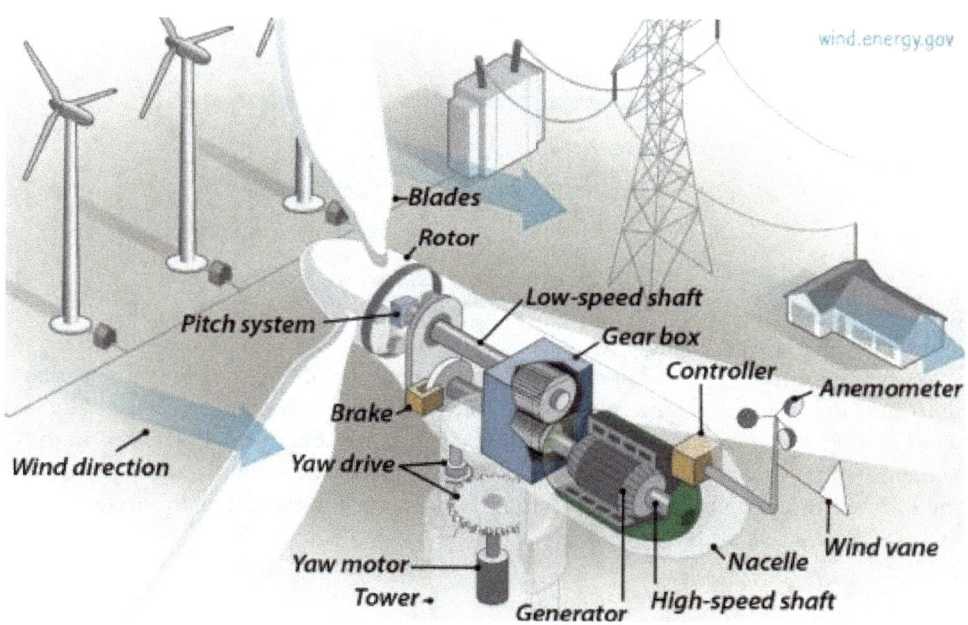

Towers: The towers are the backbone/support of the whole structure. They support the blades and nacelle structure, which houses the shaft, gearbox, generator, and controls. The tower also houses ladders or elevators for crew members to climb up and down as well as electrical cables to transport electricity from the generator to the transformer down below. Because wind speed increases with height, taller towers enable turbines to capture more wind and generate more electricity. For example, the widely used GE 1.5-megawatt turbine model consists of 116-ft blades atop a 212-ft tower for a total height of 328 feet. The blades sweep a vertical airspace of just under an acre. The 1.8-megawatt Vestas V90 from Denmark has

148-ft blades (sweeping more than 1.5 acres) on a 262-ft tower, totaling 410 feet. Another model being seen more in the U.S. is the 2-megawatt Gamesa G87 from Spain, with 143-ft blades (just under 1.5 acres) on a 256-ft tower, totaling 399 feet. Many existing models and new ones being introduced reach well over 600 feet in total height. Table WE01 below lists some of the commonly used tower structures.

Towers must also be extremely strong to support the equipment. For example, In the GE 1.5-megawatt model, the nacelle alone weighs more than 56 tons, the blade assembly weighs more than 36 tons and the tower itself weighs about 71 tons totaling a weight of 164 tons. The corresponding weights for the Vestas V90 are 75, 40, and 152, totaling 267 tons; and for the Gamesa G87, 72, 42, and 220, totaling 334 tons. Therefore, the supporting structure, which is the tower, must be extremely strong and is normally made of steel or concrete. The three main types of towers are tabular steel towers, concrete towers, and lattice towers.

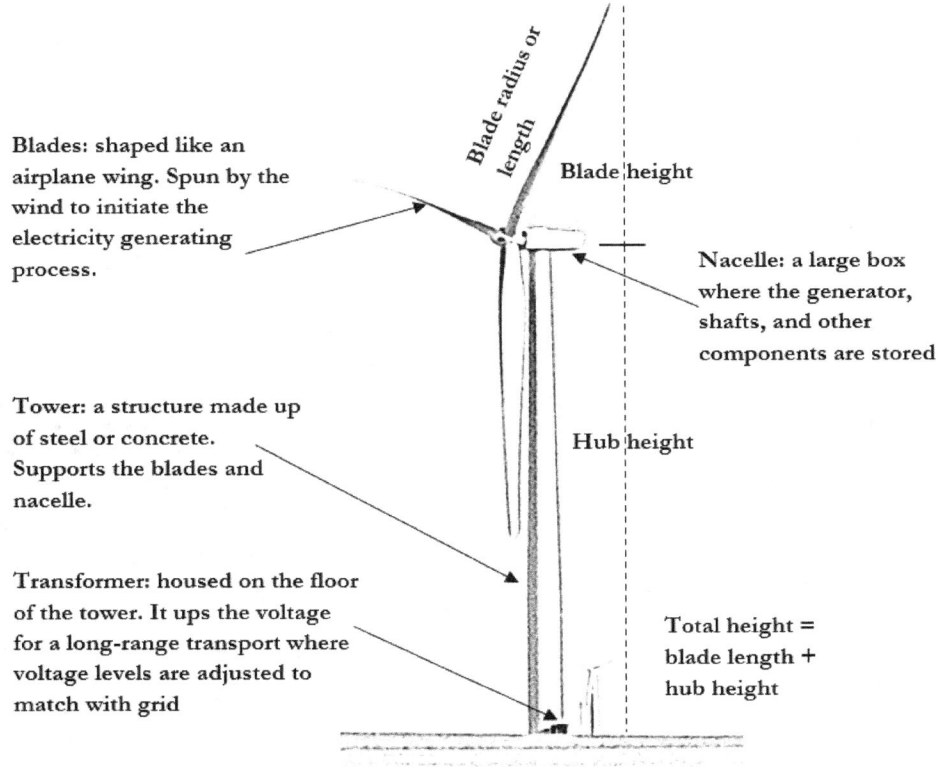

Modern-day wind turbines are massive and taller than any man-made structures. Taller towers can support bigger blades and bigger generators for higher electricity output

Table WE01: Sample of smaller and larger turbines in operation

Turbine Model	Capacity (MW)	Blade Length m, ft	Hub Height m, ft	Total Height m, ft	RPM range	Rated wind speed m/s, mph
Bonus (Siemens)	1.3	31, 102,	68, 223	99, 325	13-19	14, 31
GE 1.5s	1.5	35.25, 116	64.7, 212	99.95, 328	11-22	12, 27
Vestas V112	3	56, 184	84, 276	136, 459	6-17	12, 27
Enercon E-126	7.6	63.5, 208	135, 443	198.5, 651	5-11	
Vestas V164	9.5					

Tubular Steel Tower: The steel wind turbine tower is the most used tower type in the world. The steel tower is made in sections of around 65-130 ft (20-40m). The sections are connected with wind tower flanges. The flanges are then bolted together. All steel wind towers are in taper shape, meaning the diameter of each section decreases as the tower height increases. Because of their closed structure, the internal parts are well protected from the weather. In addition, antirust paint and coating are sprayed manually to protect the towers from corrosion, adding to the overall cost. Their large sizes also create challenges for transporting them to job sites.

Concrete Tower: Like the steel towers, the precast concrete wind turbine towers are also manufactured in sections. All the sections are transported to the site and then assembled. These concrete structures have a large carrying capacity allowing them to bear larger blades and nacelles. Since there are no fastening parts, the repair and maintenance costs are reduced largely.

Lattice Tower: Lattice wind turbine towers are made from hundreds of steel materials. They resemble traditional telecommunication towers. Many of these towers are made by companies who manufacture traditional telecommunication or electric towers since they have the experience of making these towers. Since these towers are made in smaller steel sections, transportation is easier, and the cost is lower. However, cabling and other supporting mechanisms need extra protection since they are exposed to the environment unlike tubular steel or concrete towers.

Wind turbine towers are also self-supporting (free-standing) or guyed. Most towers are self-supporting. However, there are a few exceptions depending on the site location and terrain.

Blades: The size of blades varies widely, spanning up to 150 feet wide, and resemble the wings of an airplane. Wind flows over the blades creating a lift (like the effect on airplane wings), which causes the blades to turn. Aerodynamic properties are crucial in determining how well wind turbine blades can extract energy from the wind and efficiently produce wind power.

Blades are usually fabricated from thin-walled fiber composite materials, with fiberglass-reinforced polyester or epoxy. Carbon fiber or aramid (Kevlar) is also used as reinforcement material. Fiber composite materials are high in strength and stiffness, lightweight, low density, and have superior fatigue properties. Nowadays, the possible use of wood compounds, such as wood-epoxy or wood-fiber-epoxy, is being investigated. Below are different types of fiber products used in manufacturing blades today.

Glass fibers: The stiffness of composites is determined by the stiffness of fibers and their volume content. Typically, E-glass fibers (i.e., borosilicate glass called "electric glass" or "E-glass" for its high electric resistance) are used as the main reinforcement in the composites. By increasing the volume content of fibers in the composites, the stiffness, tensile, and compression strength increase proportionally. Typically, the glass/epoxy composites for wind blades contain up to 75% glass.

Carbon fibers: Carbon fiber has more tensile strength, higher stiffness, and lower density than glass fiber. Carbon fibers are also 20% to 30% lighter than glass fibers, leading to thinner, stiffer, and lighter blades. However, because carbon fiber is more expensive, glass fiber is still dominant. Carbon fiber reinforced composites also have the disadvantage of being too sensitive to fiber misalignment and waviness, which lead to a reduction of strength.

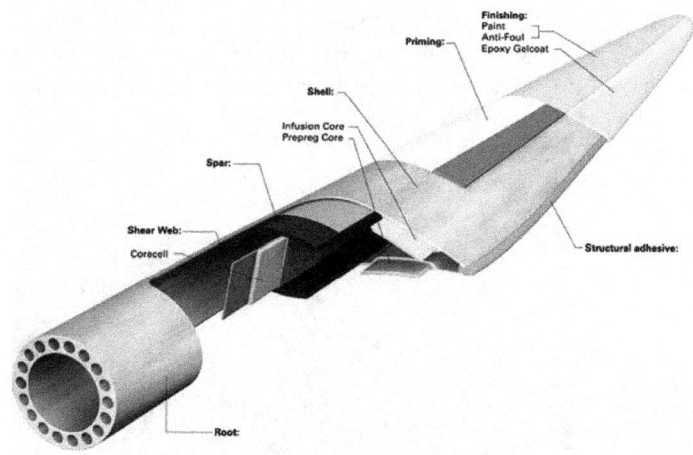

Aramid and basalt fibers: These are an alternative to non-glass, high-strength fibers. Aramid (aromatic polyamide) fibers demonstrate high mechanical strength and are tough and damage-tolerant. They are 30% stronger, 15–20% stiffer and 8–10% lighter than E-glass, and cheaper than carbon fibers. Unfortunately, they have low compressive strength and low adhesion to polymer resins. They also absorb moisture and degrade due to ultraviolet radiation. Currently, basalt fibers are used in small wind turbines and as hybrids with carbon fibers.

Hybrid composites: Hybrid reinforcements (E-glass/carbon, E-glass/aramid, etc.) are an alternative to pure glass or pure carbon reinforcements. Hybrids lead to much lower blade weight and unfortunately, to higher cost. Thus, their use is limited.

Natural fibers: Although not common, natural fibers can be used as well, especially in residential or small-scale wind projects. Natural fibers, such as sisal, flax, hemp, and jute, are low-cost, readily available and environmentally friendly. Unfortunately, they have high moisture uptake and low thermal stability leading to quality inconsistency. Another promising natural fiber is bamboo. Bamboo has high strength and durability and is also broadly available. Locally available timber is also a promising material for low-cost and reliable wooden blades.

Blade maintenance

Blades are the most vulnerable parts of a wind turbine. They are consistently exposed to extreme weather conditions, including lightning. Lightning strikes are an expected reality of a wind turbine. It is common to observe scorching damage and cracking of blades, as well as spar rupture, separation, and surface tearing in more extreme cases. Thus, all blades have a lightning protection system to reduce the effect of such strikes.

In addition, harsh weather conditions also cause significant blade damage. For example, icing on the surface of the blades under extremely low-temperature weather conditions will degrade and even stop the operation of the turbine as was seen in the Texas storm of 2021. The aerodynamics of the blade will suffer, and energy generation will be reduced. Also, the additional ice weight leads to unbalanced load distribution, which will lead to structural fatigue.

Airborne particulates also cause significant damage, especially around the tip of the blades, where velocities are higher. This damage leads to a rough surface which will degrade the aerodynamic performance of the blade and reduce power production. Therefore, continuous maintenance is a must for optimal results.

Nacelle: A nacelle is a cover housing that sits atop the tower and houses all the electricity-generating components. The housing frame is made of front and rear frame parts. The front or main frame of the nacelle is generally made of cast steel and holds the yaw system, gearbox, and the drive shaft that is connected to the blades via the hub. The rear frame is constructed of formed and welded steel and houses the generator, transformer, and electrical control cabinets.

Once the yaw system passes its rotational test and its motors are installed and pass their functional tests, the two halves of the frame are joined by heavy bolts and spring pins. The entire assembly is attached by brackets to the bottom half of the nacelle's fiberglass cover. Then the main shaft and gearbox unit along with the generator assembly are lifted into the nacelle using a gantry crane and bracketed to the gondola. Several OEMs install the transformer inside the nacelle at this point, but most install the transformer at the base of the tower. With everything in place, fiberglass upper housing is installed to cover and protect components inside. Finished nacelles are then moved out of the factory and shipped by truck or rail to wind farms to be lifted onto towers.

Nacelle - System and Components

The functional groups and systems within a nacelle of a modern wind turbine include: the yaw system, the mechanical drive train, and the electrical systems and cabinets

Yaw System

Yaw System is the component that is responsible for the orientation of the wind turbine rotor toward the wind. It allows the turbine to head itself into the wind direction, but also keeps itself locked in that position when the wind direction is stable. There are two techniques used; both are based on a bearing, brakes, drives, and positioning system:

Yaw Roller Bearing System/Passive yaw system: This technique utilizes the wind force to adjust the orientation of the rotor into the wind. It comprises a roller-bearing connection between the tower and the nacelle. A tail fin/wind vane is mounted on the nacelle to help the rotor turn into the wind by exerting a "corrective" torque to the nacelle. Simply put, the power of the wind is responsible for the rotor rotation and the nacelle orientation. This system is commonly used for small/medium-sized wind turbines since it offers a low-cost and reliable solution.

Yaw Slide Bearing System/Active Yaw System: This technique is based on automatic signals from wind direction sensors. This system is state of the art for all modern medium and large-sized wind turbines, including offshore turbines.

Brake: Stops the rotor mechanically, electrically, or hydraulically in emergencies or if wind speed is unsafe.

Pitch: Turns (or pitches) blades out of the wind to control the rotor speed and to keep the rotor from turning in winds that are too high or too low to produce electricity.

Anemometer: A measuring device that sits atop the Nacelle. It measures the wind speed and direction and transmits the data to a controller.

Controller: Starts up the machine at a cut-in wind speed and shuts off the machine at about 55 mph to prevent turbine damage at higher wind speeds. Also, it uses the data from the Anemometer to rotate the rotor/blades to face maximum wind direction.

Mechanical drive train

The mechanical drivetrain is the "powerhouse" of a wind turbine. It contains the gear-box and the generator, which are responsible for converting the rotation of the blades into electricity.

The gear-box speeds up the slow rotation of the rotor from around 5-15 rotations per minute (rpm) to higher speeds of 1,000–1,800 rpm needed to generate electricity. This is called Gearbox Drive. The other alternative is Direct Drive. Direct-drive generators can produce electricity at much lower speeds. They are directly attached to the rotor and do not require a gear-box. However, their makeup requires heavy, rare-earth materials such as neodymium and dysprosium, which makes them expensive. These are mainly used in larger offshore turbines.

Rotor: The blades and hub together form the rotor, which is attached to a low-speed shaft, which is a metal rod connected to the blades on one end and a gearbox on the other end. The shaft is spanned by rotating blades.

Gearbox: Connects the low-speed shaft to the high-speed shaft to increase the rotational speeds to a required rpm of about 1,000-1,800. The gearbox is one of the more costly and heavily weighted parts of the wind turbine. There is ongoing research and development to reduce the cost of previously discussed "Direct Drive" generators and eliminate gearboxes from the equation.

High-speed shaft: Drives the generator.

Generator: The part responsible for producing electricity.

Electrical Systems and Cabinets

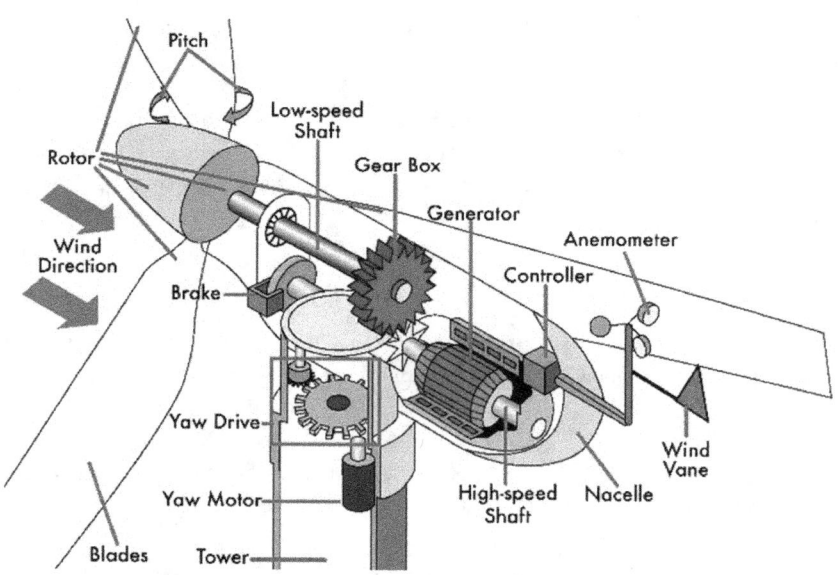

The Electrical Systems and Cabinets convert the electricity of the generator into a suitable format to match the Electric Grid, frequency AC/DC.

Transformers: Finally, all the electricity generated by the wind turbine and the generator is carried down to a transmission substation via electrical wires. The electricity is then converted into extremely high voltage, between 155,000 and

765,000 volts, for long-distance transmission on the grid. The grid comprises a series of power lines that connect the power sources to demand centers.

Wind turbine installation
(left) tower base or foundation preparation and erecting the tower in sections
(middle) installation of rotor hub and nacelle
(right) attaching the blades to the hub

Wind turbine types

There are two basic types of wind turbines: Horizontal axis turbines and Vertical axis turbines. Most wind turbines manufactured today are horizontal axis with three blades

Horizontal axis: They have three blades that resemble propeller airplane engines. The largest horizontal-axis turbines are as tall as 20-story buildings and have blades more than 100 feet long. Longer blades on taller towers generate more electricity. Nearly all wind turbines currently in use are horizontal-axis turbines.

Vertical-axis: The blades are attached to the top and the bottom of a vertical rotor and have the look of a giant two-blade eggbeater. Some versions of the vertical-axis turbine is 100 feet tall and 50 feet wide. Very few vertical-axis wind turbines are in use today because they do not perform as well as horizontal-axis turbines.

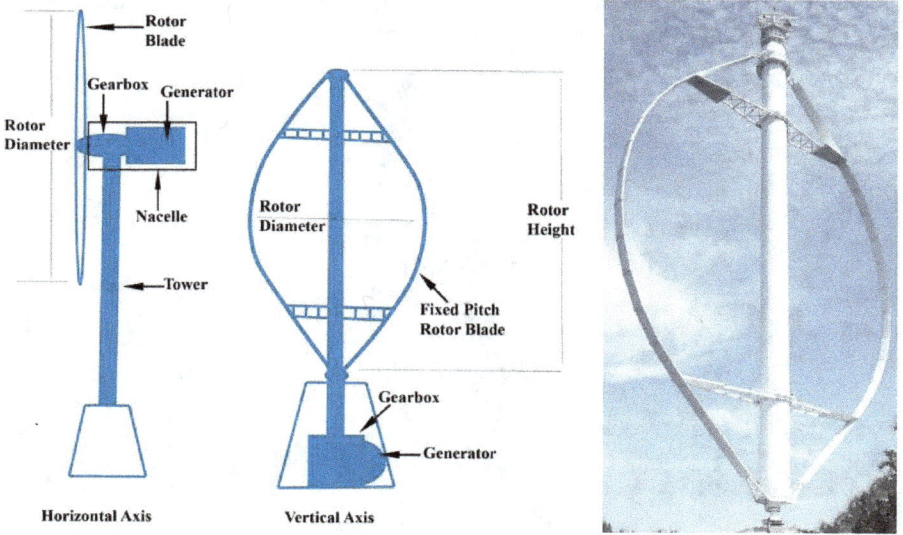

Wind farms

Wind farms are turbines grouped over a large territory. Wind farms provide power to the electricity grid of large-scale utility companies. They can either be onshore (on land), like most projects today, or offshore (on oceans/seas), which is gaining acceptance rapidly.

Onshore wind farms are wind turbines installed on land and thus, installation costs are lower. Still, installation is extremely challenging and requires heavy-duty machinery due to the tower, turbine, and nacelle size and weight. Below is a pictorial illustration of an installation process for a single turbine.

To provide an uninterrupted flow of air ensuring power generation, onshore wind turbines require a minimum distance of 150m from any obstructions and a turbine-to-turbine separation of 7 times the diameter of the rotor. Therefore, wind farms require a large acreage of land space. For example, the largest onshore wind farm in the U.S. (Alta wind farm, California) has about 600 turbines generating over 1.55 GW of electricity. The farm was constructed in 2010 at $2.9 Billion and covers an area of 3,200 acres (5 square miles, mi2) of land. Fortunately, energy generated from onshore wind farms can easily be added to the grid. The maintenance cost is reasonably low as well. In some cases, investment payback can be as fast as two

Onshore wind farms are a great way of utilizing deserts and wastelands

years. For this reason, most wind farms in operation today are onshore. Onshore wind energy continues to expand and global capacity from onshore wind energy is projected to reach nearly 750 GW by 2022. Today, onshore wind energy is the 2nd cheapest source of energy behind solar energy at a cost of below 5 cents per kWh. As a reminder, the fossil fuel and nuclear energy fueled cost in the U.S. today is about 12 cents per kWh for residential and 7 cents per kWh for industrial.

Site location for wind farm projects requires careful planning that extends beyond merely erecting wind turbines in a windy area. A detailed study of the site, based on topology and wind blow characteristics (how fast or slow and how often the wind blows), must be studied ahead of time. After all, wind speeds generally change throughout the day and from season to season. For example, in parts of California, where numerous wind turbines are located, the wind blows more frequently from April through October, and the wind is usually strongest in the afternoon. In Montana, strong winter winds channeled through Rocky Mountain valleys create more intense winds during the winter.

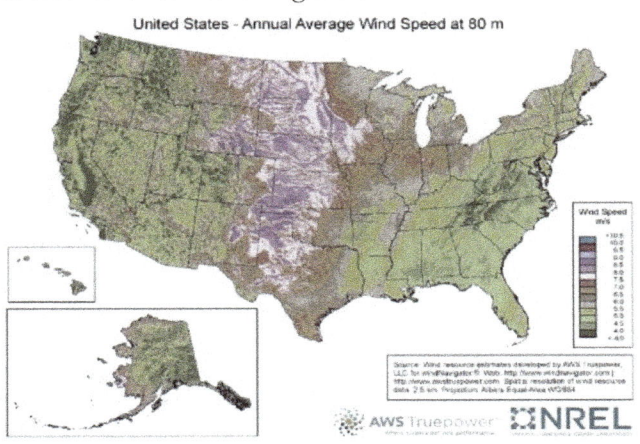

The Midwest regions of the United States has a consistent wind speed throughout the year. Unsurprisingly, states in the Midwest, such as Texas, Oklahoma and Kansas are leaders in wind power capacity.

The right places for wind turbines are where the annual average wind speed is at least 9 mph for small wind turbines and 13 mph for utility-scale turbines. As previously mentioned, wind speeds over 55 mph cause more harm than good. At these low required speeds, every region has the potential to produce wind energy. Favorable sites are areas with consistent wind resources such as tops of smooth and rounded hills, open plains, oceans/seas, and mountain gaps that funnel and intensify wind. In addition, since higher elevation produces more energy, wind turbines are also placed on towers that range from about 500 feet to as much as 900 feet tall.

The U.S. accounts for five of the top ten onshore wind farms in the world. The largest five wind farms in the U.S. are as follows:

1. Alta Wind Energy Centre (3rd largest in the world)
Located in Tehachapi, Kern County, California, the Alta Wind Energy Centre is the biggest wind farm in the US. With a combined installed capacity of about 1.55 GW, it has about 600 turbines and covers over 5 mi2 land area. It was commissioned in 2010 at $2.9 billion.

2. Los Vientos Wind Farm
With an installed capacity of 910 MW, the Los Vientos Wind Farm in Texas was completed in five phases starting in 2012 and ending in 2016. The combined five sections of the farm have about 426 installed turbines.

3. Shepherds Flat Wind Farm
Located near Arlington in Eastern Oregon, the Shepherds Flat Wind Farm is the third largest wind farm in the U.S. with an installed capacity of 845 MW. The wind farm is spread over 30 square miles of land area. Construction on the wind farm started in 2009, with an estimated cost of $2 billion. The Shepherds Flat Wind Farm began commercial operations in 2012. It consists of 338 GE2.5XL turbines, each with a rated capacity of 2.5MW.

4. Roscoe Wind Farm
The 781 MW Roscoe Wind Farm near Abilene, Texas, is the fourth biggest wind farm in the U.S. The wind farm was built in four phases between 2007 and 2009 for a total of 627 wind turbines.

5. Horse Hollow Wind Energy Centre
Located in Taylor and Nolan County, Texas, the Horse Hollow Wind Energy Centre has an installed capacity of 735.5 MW. The wind farm was commissioned in

four phases during 2005 and 2006. At the time of its completion in 2006, it was the largest wind farm in the world. The farm has about 421 wind turbines generating electricity for nearly 180,000 Texan households.

Jiuquan Wind Power Base/Gansu Wind Farm, China
For comparison purposes, the largest wind farm in the world, Jiuquan Wind Power Base in China, has an installed capacity of about 8 GW. The wind farm construction began in 2009. When completed, it is projected to have 7,000 wind turbines for a total installed capacity of 20 GW at a cost of nearly $17 billion.

Offshore wind farms, on the other hand, are built offshore to take advantage of the higher and consistent wind flow generated in the oceans. On average wind flows are 20% higher in the oceans compared to onshore. Thus, offshore wind farms generate more power at a more stable rate. Additionally, offshore wind farms have the advantage of large open spaces for installation. Thus, larger turbines, some twice or three times larger than onshore turbines, can be installed. The larger turbines, along with a more consistently higher wind flow rate allow for greater power output at a higher efficiency and lower operations cost.

Although the operation is similar to onshore wind power, the installation is much more demanding. Building a tower foundation on ocean floors and installing the rest of the turbine components require advanced techniques and increased financial commitment. Offshore wind farms also require platforms, underwater cables and interconnection, and other factors that lead to a nearly 20% increased installation cost. Despite the challenges of installation cost, the advantages of limitless offshore space, stable wind patterns, and declining installation costs make offshore wind farms far more beneficial. In fact, offshore wind power is projected to grow at a rate of 16% between 2019 and 2030, reaching a cumulative capacity of 142 GW by 2030. As of 2018, offshore wind power had a global capacity of 23 GW.

Globally, both the European Union and China are making great strides in expanding offshore wind farms. Unfortunately, there is no offshore wind farm to speak of for the United States. The largest offshore wind farm in the U.S. is Block Island Wind Farm, located off the coast of Rhode Island in the Atlantic Ocean. The farm has five turbines installed in 2016 for a total capacity of 30 MW. By comparison, the United Kingdom (U.K.) has some of the largest offshore wind farms in the world ranging from 500MW to Over 1GW of power capacity. Some of the largest offshore wind farms in the world include:

1. Hornsea

The Hornsea-1 offshore wind farm, with an operating capacity of 1.2 GW is currently the world's largest offshore wind farm in the world. The wind farm consists of 174 turbines covering an area of 157 mi2 (407 km2) of the North Sea off the coast of the U.K. The project is the first offshore wind farm to generate over 1 GW of electricity. According to the official project website, the farm generates enough power to meet the energy needs of over a million British households. Hornesea-1 is the first of a planned four-phase project for a total capacity of 6 GW of electricity when completed.

2. Walney Extension

Walney Extension of England's Walney Island is the second-largest offshore wind energy farm in the world with an operational capacity of 659 MW. The operation began in 2018 with 87 turbines, enough to generate clean electricity to power 600,000 homes.

3. London Array

Third on the list is the London Array offshore wind farm with a capacity of 630 MW located off the coast of the U.K. The plant was commissioned in April 2013 with 175 turbines.

4. Gemini Wind

The only top five project located outside the U.K. waters is the Gemini wind farm in the Dutch part of the North Sea. With a generational capacity of 600 MW, the project generates electricity for approximately 785,000 households. The plant was officially commissioned in 2017 and featured 150 turbines.

5. Beatrice

Finally, the Beatrice offshore wind farm, off the coast of Scotland is one of the newest projects. The project was commissioned in 2019 with a capacity of 588 MW. The farm features 84 wind turbines and generates enough electricity for over 450,000 homes annually.

Upcoming Project: The world's largest offshore wind project, Dogger Bank Farm, in the North Sea off the coast of the U.K. is to be constructed in three phases for a total 3.6 GW capacity. According to the project website, the project will generate electricity for nearly 6 million homes in the U.K. or 5% of the U.K. electricity demand when completed in 2026. The project will use the world's largest, 13 MW General Electric Haliade-X, turbines. The turbines stand more than 850 feet tall and are designed to generate 45% more power than current best-in-class offshore turbines. According to GE, the turbine's blades measure longer than a soccer field. A single rotation can supply enough electricity to power the average British household for two days.

Small-scale wind power

Larger wind farms are more efficient and effective. However, wind farms require greater financial investment, resources, and vast areas of open space under the right wind conditions. Therefore, they are only efficient and effective at the utility-scale. For the smaller clients, such as businesses, farms, and residentials, small-scale wind turbines offer an opportunity to benefit from wind energy. Small-scale turbines are deployed in a wide range of locations at an elevation between 30 ft to 140 ft. By comparison, utility scale turbines operate at elevations between 300 ft to 900 ft. One advantage of small-scale turbines is that they can operate at low wind speeds of only 9 mph. Another advantage is their reduced cost and versatility make them preferable for on-site production, minimizing the need for transmitting energy over the electric grid from central power plants or wind farms. However,  because wind speed is slower and inconsistent at low elevations, small-scale turbines generate much lower power.

Small-scale wind power utilizes two different types of turbines:

Horizontal axis wind turbine (HAWT): Recent advances have improved many of the features of these turbines. Newer models can generate more power with fewer rotations per minute (RPMs) making them more efficient. Some even have curved blades to help reduce noise levels. Additionally, new models use rare earth magnets in their generators, allowing for smaller and lighter generators. The rotors are also equipped with brakes or pitched blades to protect the turbines against damage from high winds.

Vertical axis wind turbine (VAWT):
These turbines are designed with blades that rotate around a vertical shaft. These turbines operate at lower wind speeds thus, can be placed at a lower elevation than HAWTs. The shape of their blades also allows them to generate power from wind blowing in any direction, including vertically. VAWTs also have smaller space requirements and can be placed closer together than HAWTs. Since the generators are located at ground level at the bottom of the shaft, maintenance is also much easier.

However, VAWTs generate less power than their counterparts HAWTs because they operate at lower wind speeds. Also, the lack of a brake system makes the blades more vulnerable to damage by high winds. In short, HAWTs are far more common due to their superior efficiency and generation capabilities. However, VAWTs may be preferred in locations with less available space or where wind speed and direction are inconsistent.

Small-scale wind power can be used in an off-grid or on-grid system. An off-grid system is where wind turbines are not connected to the electrical transmission grid. These are generally installed in areas far from the grid and/or in places that are expensive to connect to the grid. Off-grid systems use direct current (DC) to power devices in remote locations, such as telecommunications equipment and water pumps in rural areas. These systems can also use battery storage to provide backup power when the wind is not blowing.

On the other hand, on-grid wind power is where the system is connected to the grid. Many newer turbines designed for on-grid systems have built-in inverters to convert the electricity to Alternating Current (AC) for compatibility with household appliances. An added benefit of the on-grid system is that any excess electricity generated can be put back into the grid and sold to the utility company for profit.

A modern building with four small wind turbines (HAWT) on a rooftop. Power from these small-scale turbines contributes to the overall electricity demand for the building.

Active Solar Energy

The sun is the ultimate source of energy. It is the source of the fossil fuel we have used to build our economies for centuries. It is also the source of renewable energy we can use to build our economies now and in the future. According to the National Renewable Energy Laboratory (NREL), the sun produces enough energy every second to cover the earth's power needs for 500,000 years. The energy that reaches the earth in one hour is enough to meet our energy demand for a year. As previously discussed, we have used energy from the sun in passive solar design for thousands of years. Recently, we have learned to use the same energy in much more direct and effective ways, known as Active Solar Energy.

Active Solar Energy is the concept of harnessing radiant light and heat from the sun to generate electricity as well as heating for residential, commercial, and industrial spaces. It was first used commercially in the mid-1800s and has been growing rapidly in response to a growing demand for cleaner and sustainable energy. Active solar energy is harnessed using two different techniques: Photovoltaic system (PV) and Thermal energy.

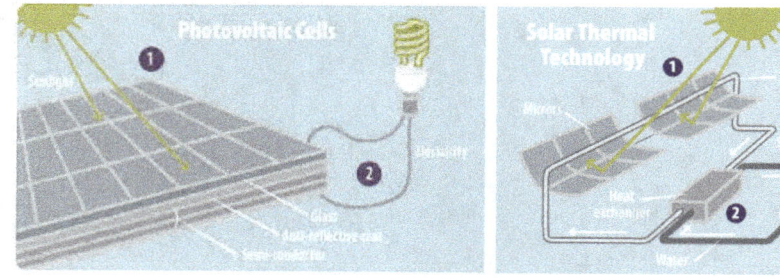

source: U.S. Environmental Protection Agency

A photovoltaic system (PV) commonly known as solar power, is the process of converting light radiated from the sun directly into electricity using solar panels and additional hardware. Thermal energy, on the other hand, requires using heat from the sun to boil water or other fluids and using the resulting steam to spin a turbine and a generator. According to the U.S. Energy Information Administration (eia), the U.S. had over 125 GW of installed solar power capacity at the end of 2023, enough to power 30 million homes. The progress is impressive compared to the 10 GW of solar power capacity ten years ago. Today, solar power accounts for 3.4% of total U.S. electricity generation and has ranked either first or second in capacity added to the entire U.S. electric grid system every year since 2013. In 2023 alone, over 30 GW of solar power capacity was added to the U.S. grid.

Globally, solar power accounts for 4.5% of all electricity generated. Many countries and territories have installed significant solar power capacity into their electrical grids to add to their conventional energy sources. The worldwide growth of photovoltaics varies enormously by country. Combined, the installed solar power capacity in the world had reached north of 1.2 TW (1,200 GW) at the end of 2023. The growth is remarkable compared to the 140 GW global capacity ten years ago. As of 2023, there were at least 47 countries worldwide with a cumulative PV capacity of more than one GW. China is the leader accounting for one-third of the total global solar power capacity followed by the U.S., Japan, Germany, and India. The top five solar power-producing countries in the world account for 60% of all global capacity.

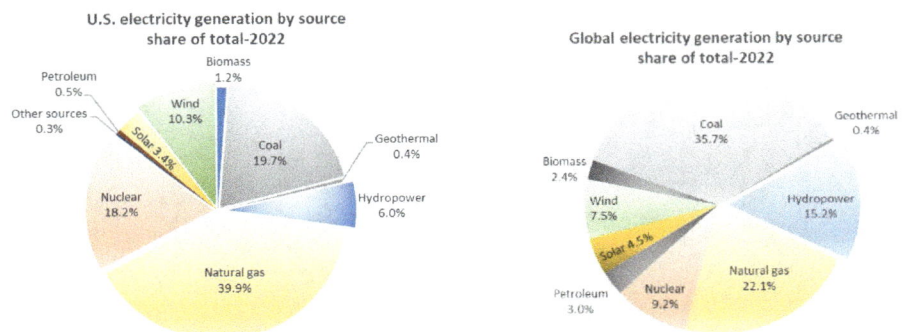

Solar power accounts for 3.4% of the U.S. and 4.5% of global electric grid

One reason for the rapid growth in solar power capacity is technological advances that have improved panel efficiency from 5% in the 1950s to nearly 30% today. In addition, favorable government policies and incentives have contributed greatly to the advancement. For example, the U.S. Federal program, known as the "Investment tax credit for solar power (ITC)", was enacted in 2005 and extended several times to 2021. The program offered tax credits and other incentives to lower the cost of solar power installation and allowed more businesses and communities to participate. Improved technology and encouraging policies contributed to an astonishing 89% decline in solar energy costs in the last ten years in the United States. According to the SEIA, the average-sized residential system cost has dropped from a pre-incentive price of $40,000 in 2010 to roughly $20,000 today. Additional investment was allocated for the renewable energy industry, including solar power, as part of the Inflation Reduction Act (IRA) in 2023. The impact of the IRA on the industry remains to be seen.

The cost of generating electricity from renewable sources, such as hydropower, geothermal, and bioenergy has always been competitive with fossil fuel-based utilities. However, the increasing competitiveness of solar and wind power against other technologies has been impressive. The cost of electricity from solar power has been reduced by 89%, from 45 cents/kWh to 5 cents/kWh in just a decade. Similar progress has been made for wind power, reducing the cost from 11 cents a decade ago to 3 cents today. Currently, solar and wind power are the cheapest sources of electricity in the U.S. By comparison, the cost of electricity from coal, natural gas, and nuclear facilities is 11 cents, 6 cents, and 16 cents per kWh respectively.

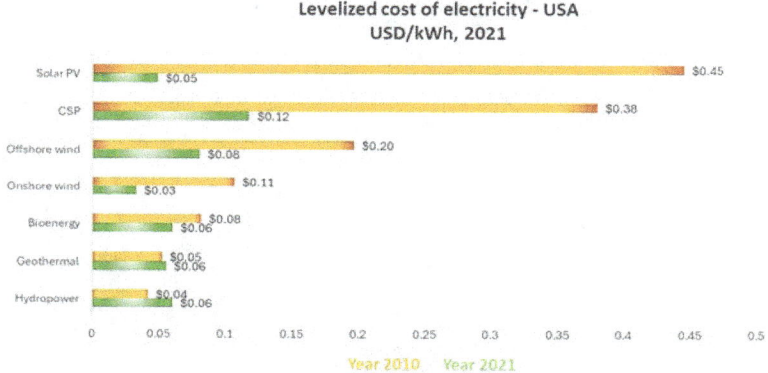

(above) levelized cost of electricity from renewables in 2010 vs 2021
(bottom) levelized cost of electricity from all sources – US, 2021

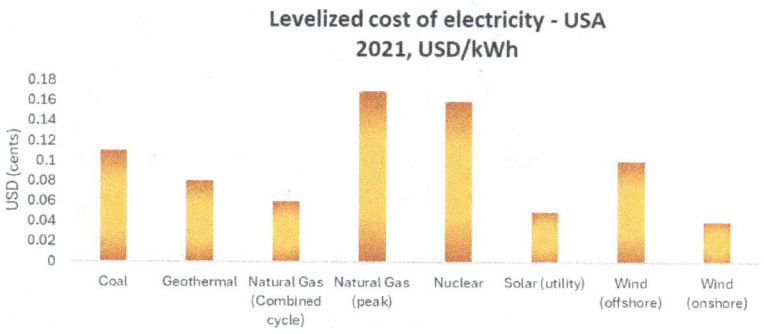

Economically, the solar power industry has made a significant contribution to the labor force as well. According to US Energy & Employment Report 2022, the energy sector contributed nearly 8 million jobs or 4.5% to the overall U.S. job market in 2021. Solar power-related jobs accounted for 39% of the energy sector jobs employing 333,887 workers, more than the coal, oil, and natural gas industries combined. The solar photovoltaic installer is also the fastest-growing job in America today.

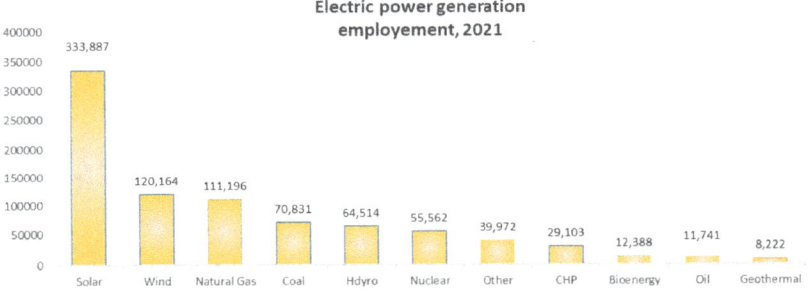

More importantly, the solar power-related jobs are spread out over nearly all 50 U.S. states. As previously mentioned, renewable energy allows each region to develop an energy portfolio mix based on its natural climate. Unsurprisingly, regions with the hottest climate lead in solar power consumption and employment opportunities. For example, according to the Solar Energy Industries Association (SEIA) 2022 report, the State of California accounts for about 32% of U.S. solar power capacity. The state accounts for over 30% of all solar power jobs as well. In fact, the top ten solar-producing states accounted for over 80% of all U.S. solar power production in 2021

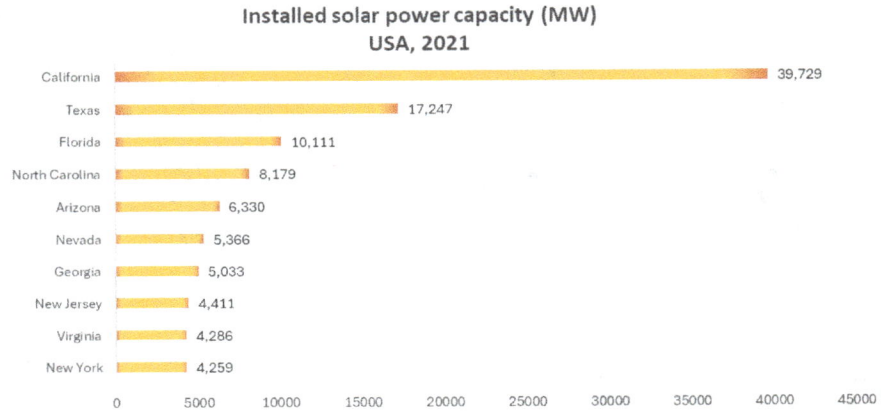

Globally, China outpaces any nation with nearly 400 GW of total solar power capacity installed.

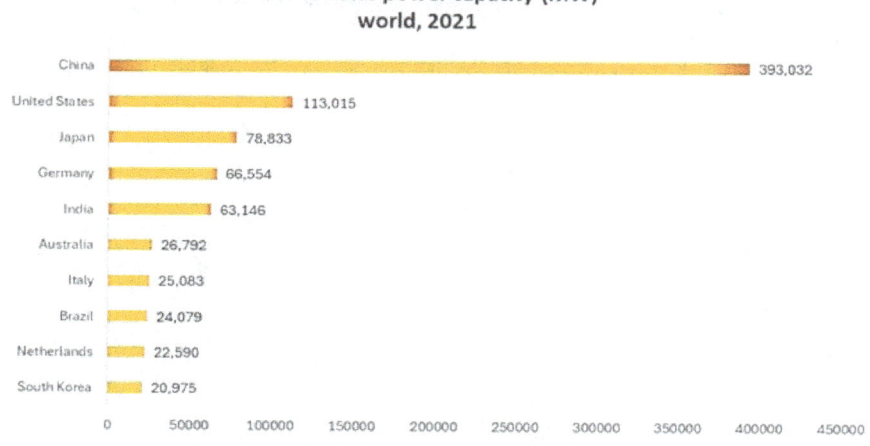

Types of solar panels

Powering small devices, such as calculators, watches, and other small electronic devices with small photovoltaic (PV) cells, was generally common knowledge for a very long time. However, photovoltaic effect, which is the science of generating electricity with solar cells, was first discovered in 1839 by Edmond Becquerel. Bell Lab's development of the first silicon PV cell in 1954 elevated the technology and allowed PV cells to power larger devices over extended periods. Photovoltaic (PV) gets its origin from (photon), which is the process of

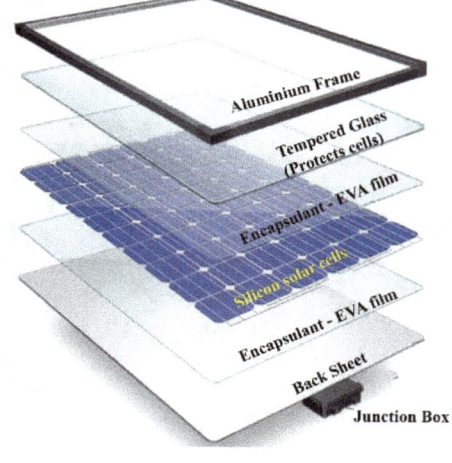

Silicon Solar Panel

converting light to (voltage), which is electricity. Thus, photovoltaic effect is the process of converting light into electricity. This process requires multiple hardware, of which solar cells, made from semiconductor materials, are the critical component. Numerous solar cells put together are called solar panels or solar modules. In addition to solar cells, solar panels consist of metal frames and glass casing units for surface protection. They also consist of wiring cables to transfer the electric current from the semiconducting cells to the devices to be powered. Further development led to the arrangement of multiple panels, put together in arrays to generate electricity on a larger scale.

Silicon solar panels are the most used solar panels, occupying more than 90% of the global PV market. These panels vary in their manufacturing process, appearance, performance, and costs. They have an efficiency (the rate at which the solar cell converts sunlight into electricity) of up to 33%. As the name indicates, silicon solar panels are made from silicon, which is the most common semiconductor material used today. Silicon is one of the most abundant nonmetal elements. In fact, more than 90% of the Earth's crust comprises silicate minerals, making silicon the second most abundant element in the Earth's crust after oxygen.

Silicon has the second highest melting and boiling points of the non-metals. Thus, silicon-based panels can withstand harsh elements, whether it be from the sun's heat or freezing temperatures. More importantly, exposing silicon to the sun breaks its molecular bond leading to loosened protons and electronics. The loosened electrons are then captured and used as electricity. Silicon solar panels are basically silicon cells covered with a glass sheet and framed together. Putting the cells in panels makes transporting and installing them much more manageable. Solar panels are installed on roofs of residential or commercial buildings or deployed on ground-mounted racks to create massive, utility-scale systems. Solar panels come in standardized 60, 72, and 96-cell variances.

Silicon cells are made from either monocrystalline or polycrystalline silicon crystals. **Monocrystalline** solar cells are manufactured from a single silicon crystal. Therefore, they have the highest efficiency and power capacity. They can reach efficiencies of higher than 20 percent and have a power capacity of more than 400 watts (W) and few exceeding 500 W. Monocrystalline cells are distinguished from others by their dark color, which is the appearance of pure silicon crystals when exposed to sunlight.

Manufacturing monocrystalline cells is an energy-intensive process and results in a waste of fragmented silicon material. The process leads to higher costs. Fortunately, the silicon waste is processed to manufacture other cells, referred to as polycrystalline solar cells.

Polycrystalline solar cells are composed of fragments of silicon crystals melted together in a mold before being cut into wafers. Thus, they are the least efficient and least expensive. These panels usually have efficiencies of around 17 percent and up to 300 W power capacity. Polycrystalline panels appear bluish due to how light reflects off the cell's silicon fragments.

Another commonly used photovoltaic panel technology is **thin-film solar cells.** These cells are made from thin layers of semiconductor material, most commonly cadmium telluride (CdTe). Manufacturers make this type of thin-film cells by placing a layer of CdTe between transparent conducting layers that help capture sunlight. The cells are also layered with a flexible glass on top for protection.

Thin-film solar panels are also made from amorphous silicon (a-Si), a material similar to the composition of monocrystalline and polycrystalline panels. Although these panels use silicon in their composition, they are not made up of solid silicon wafers. Instead, they're composed of non-crystalline silicon placed on top of glass, plastic, or metal. Lastly, Copper Indium Gallium Selenide (CIGS) panels are another popular type of thin-film technology. Electrodes are placed on the front and the back of the material to capture electrical currents.

Overall, thin-film solar cells are flexible and lightweight making them ideal for portable applications, such as backpacks, RVs, and boats. Their lighter weight and flexibility contribute to a lessor, labor-intensive installation leading to a cheaper installation cost. These cells, however, have the least efficiency at about 11 percent and thus, are the least expensive. Thin-film panels do not come in uniform sizes, unlike monocrystalline and polycrystalline solar panels. As such, the power capacity from one thin-film panel to another will largely depend on its physical size. Also, because of the variation of materials used to make these cells, thin-film solar panels have a bluish and blackish color.

Bifacial solar panels are another type of solar panels available in the market. These panels capture sunlight from both the front and back of the panel, thus producing more electricity than comparably sized, traditional solar panels. Bifacial panels have a 10% to 30% higher output than monofacial panels. The increased production capacity is reflected in their higher purchasing cost. Bifacial solar panels are typically manufactured with monocrystalline solar cells, but polycrystalline bifacial solar panels exist as well.

Unfortunately, installation has its challenges. Bifacial modules mounted flush on a rooftop block any reflected light from reaching the backside of the cells. Thus, they have to be installed at a tilt. The higher the tilt, the more power they can produce. For that reason, ground-mounted installation is preferred because there is more room for tilt allowing light to reflect on both sides of the panel.

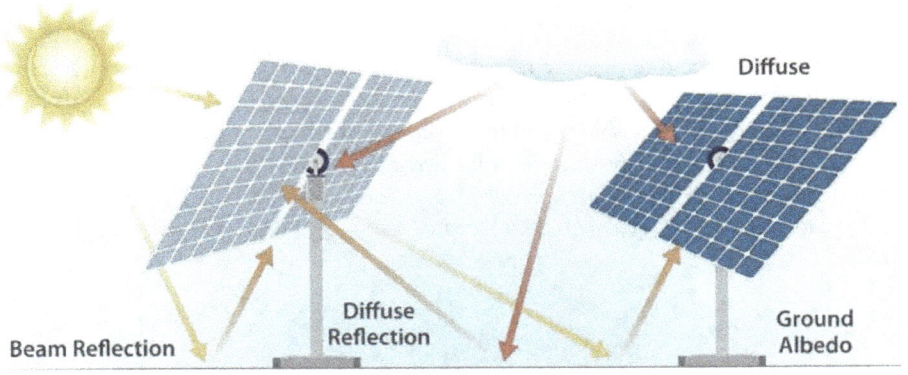

Bifacial solar panels
source: **National Renewable Energy Laboratory (NREL)**

III-V solar panels are mainly constructed from Group III elements; gallium and indium, and Group V; arsenic and antimony of the periodic table. These solar cells are generally much more expensive to manufacture than other technologies. But they convert sunlight into electricity at a much higher efficiency of up to 47%. Because of their high efficiency rate, these solar cells are often used on high-end projects, such as space satellites, unmanned aerial vehicles, and other applications that require a high ratio of power to weight.

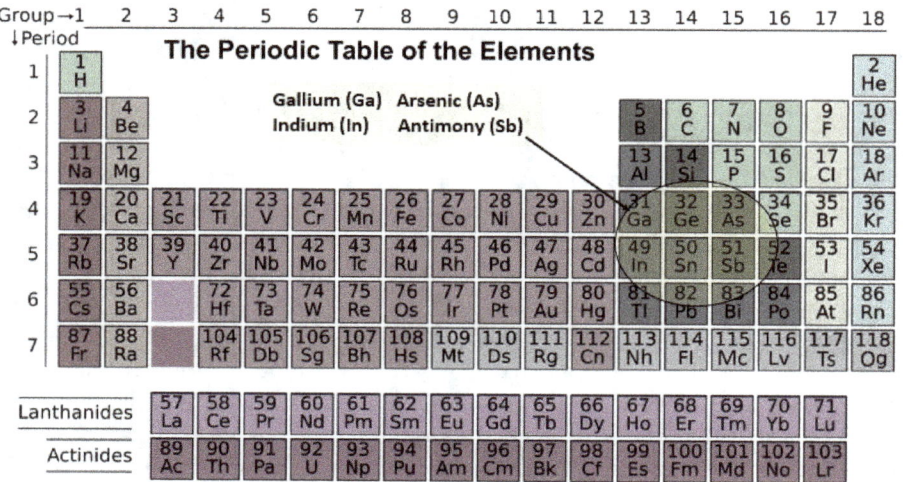

Next-generation solar cells include those that are at the research and experimental level. There is an ongoing effort worldwide to develop new photovoltaic technologies and materials to further improve panel efficiency and reduce cost. Research materials include organic materials, quantum dots, and hybrid organic-inorganic materials. In addition to promising newer solar panels, continuously improving the already available panel technology contributes to a more reliable, efficient, and cheaper energy source. Solar power is already one of the cheapest sources for generating electricity. According to the International Energy Agency (IEA), solar power is also expected to become the world's biggest source of electricity by 2050.

Additional hardware

In addition to solar panels, generating solar power by photovoltaic requires other hardware, including inverters, charge controllers, batteries, wirings, and framing.

Inverters are additional hardware connected to solar panels by copper wiring. Their purpose is to convert the direct current (DC) generated by the panels into usable alternating current (AC). Inverters are also used for monitoring and optimizing the solar power system. Thus, they serve the purpose of an online communication portal. They provide system information on the amount of solar energy produced. They also offer an inside look into the system to ensure correct functionality. There are three different solar power inverter technologies: microinverters, macroinverters, and power optimizer inverters.

Microinverters are attached to each panel and each panel is monitored and optimized separately. The energy produced is converted to AC at each panel rather than sent to a central location. Thus, a failure in a single panel does not impact the other panels. Microinverters are the most efficient and thus, the most expensive inverters. Adding to their cost are also the challenges of maintenance or repair because of the many different locations where these inverters are installed.

Macroinverters/String Inverters are used in a centralized way rather than the individualistic approach of the microinverters. In this system, each panel is wired together into a string, and multiple strings are connected to a single central inverter, usually located at ground level. The central inverter will convert all the electricity from the solar panels into AC. Thus, a failure in one panel can impact the performance of the overall array of panels. The output is as efficient as the least productive panel. Macroinverters are preferred by most small-scale solar power projects due to their affordability. Maintenance and repair are also less challenging since there is only one central location, mostly at ground level.

Power Optimizer Inverters combine both technologies of micro and macro inverters. They are considered as efficient as microinverters but at a slightly cheaper cost. Like microinverters, power optimizers are located on each solar panel. This concept helps improve the efficiency of the panels. However, like macroinverters, the energy is sent and converted to AC at a centralized location, reducing maintenance and repair costs.

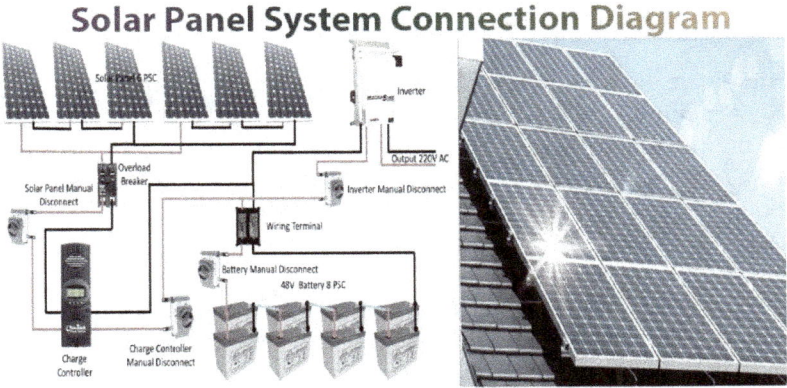

Charge controllers are a necessity for a battery-backed solar power system where excess power is stored in a battery for future use. They are the "on and off" switches for a battery system, allowing power to pass when the battery needs charging and cutting it off when the battery is fully charged. Charge controllers also contribute to the life span of the battery by improving charge quality. They prevent the battery from being overcharged. They also prevent the battery from discharging in the absence of sunlight.

Batteries: Solar panels can only produce power when the sun is shining. Even then, the amount of sunlight varies depending on location, time of day, the season of the year, and weather conditions. Thus, storing excess solar power is increasingly becoming important. Batteries provide storage and thus, provide energy reliability. They can also be arranged in many different arrays to provide storage for all scales of solar panel installations. Currently, less than 5% of solar power systems are paired with battery storage. It is estimated that the number will grow to more than 25% by 2025 because of the growing demand from homeowners and businesses.

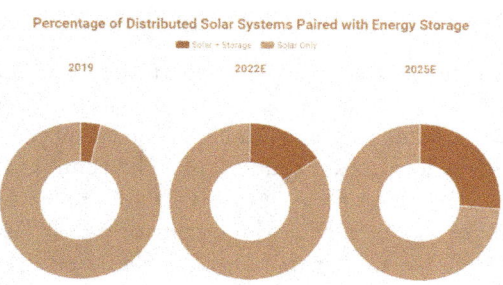

Solar power installations

Solar PVs are installed on various scales including residential, commercial, community, and utility. At the utility level, however, solar power generates the cheapest electricity of any renewable or fossil fuel-based power.

Residential-scale solar power: typically installed on rooftops of homes or in open land (ground-mounted) and are generally between 5 and 20 kilowatts (KW), depending on the property's size. The average U.S. residential solar installation is about 5 KW or around 20 panels. Unfortunately, the average cost of solar power installation in an average U.S. residence is $20,000. In addition to the cost of financing or leasing, it can take decades to recoup the benefits.

Commercial-scale solar power: generally installed at a greater scale than residential solar. Though individual installations can vary significantly in size, commercial-scale solar power provides on-site power to businesses. Data from SEIA's annual "Solar Means Business" report show that major U.S. corporations, including Apple, Amazon, Target, and Walmart, are investing in solar energy at an incredible rate. Through 2018, the top corporate solar power users in the U.S. have installed more than 7 GW of capacity across the country in more than 35,000 different facilities. More than half of the 7 GW of corporate solar capacity has been installed in the last three years.

Both residential-scale and commercial-scale projects can be installed as off-grid or on-grid systems. Off-grid is where the project is set as a stand-alone system. This means the facility generates enough power for itself and is not connected to the utility company's grid. On-grid systems, on the other hand, are tied to the utility company's grid and can benefit from what is known as **Net-metering.** Net-metering is a solar incentive program that allows consumers to store excess solar power in the utility's electric grid in exchange for credits. Then, at night or other times when the solar panels are under-producing, consumers pull energy from the grid and use the credits to offset energy costs. The program saves utility companies from having to generate additional power during peak times while rewarding consumers for their excess solar power.

Community-scale solar power is a viable solar option for solar enthusiasts who cannot or choose not to install solar panels on their property. Community-scale solar projects are typically built in a central location allowing residential consumers to subscribe and receive many of the benefits of solar power without installing solar panels on their property.

Utility-scale solar power: These projects are typically large installations that provide solar power to many utility customers. For example, one of the largest solar farms in the United States, Solar Star, produces 579MW using 1.7 million solar panels over 3,200 acreages of land space. Most utility-scale solar projects are installed on land, with few exceptions where they are installed on water or sea surface. Solar projects on water surface are known as floating solar farms and provide an alternative in areas where land space is limited, such as in Singapore.

Environmentally, solar power is a clean source of energy. According to the Environmental Protection Agency (EPA)'s Greenhouse Gas Equivalencies Calculator, the average American home going solar for a year is like:

- Reducing carbon dioxide emissions by more than 12,500 pounds.
- Not burning over 8,000 pounds of coal.
- Driving about 18,000 miles less.
- Not charging 937,683 smartphones.

Solar thermal (heat) energy

The second way of utilizing active solar energy is solar thermal/heat energy. It is a process of using panels to capture the sun's heat (rather than light) and use it directly for heating. It can be utilized on a smaller scale to heat residential, commercial or industrial spaces. It can also be used on larger utility-scale projects to generate steam for electricity production.

Small-scale solar thermal energy utilizes many different technologies to collect and convert solar radiation into usable heat energy for various purposes, including heating the interior of buildings, greenhouses, and even swimming pools. For example, solar water heating systems are composed of solar collectors, insulated piping and hot water storage tanks to collect the sun's thermal energy. Then, the accumulated energy or hot water is circulated with either an electric pump (active) or using gravity (passive). Active solar water heating systems are more common in residential and commercial use. Passive solar water heating systems are typically less expensive, but they are also less efficient.

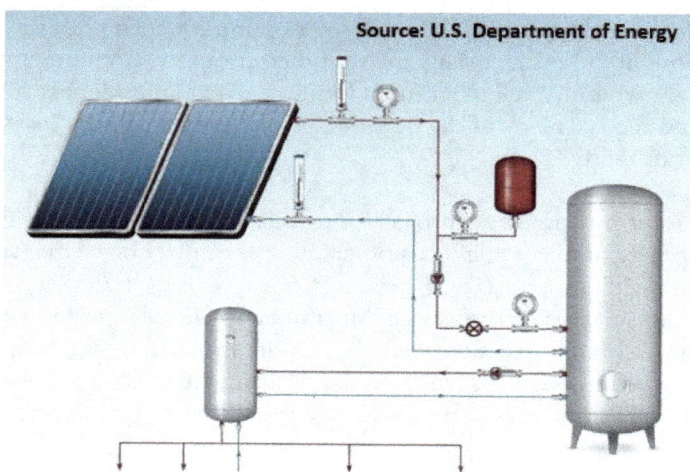

Source: U.S. Department of Energy

**Passive water heating system:
The heated water is distributed using gravity.
No electric pumps required**

Larger or utility-scale solar thermal energy is a process of using many panels to collect heat which is used to steam water or other fluids. The steam is then used to spin a large turbine and generator to produce electricity. This method is called concentrated solar power (CSP) system. The process harnesses heat from the sun to provide electricity for large power stations.

There are three main types of CSP systems: linear concentrator, dish/engine and power tower systems.

Linear concentrator systems collect the sun's energy using long rectangular, curved (U-shaped) mirrors. The mirrors are tilted toward the sun to expose attached tubes or receivers to maximum sunlight. The reflected sunlight heats a fluid flowing through the tubes. The hot fluid is then used to steam water in a conventional steam turbine generator to produce electricity. Linear concentrating collector fields consist of many collectors in parallel rows that are typically aligned in a north-south orientation to maximize annual and summer energy collection. In these systems, the collector field is oversized to heat a storage system during the day, so the additional steam it generates can be used to produce electricity in the evening or during cloudy weather.

There are two major types of linear concentrator systems: parabolic trough systems, where receiver tubes are positioned along the focal line of each parabolic mirror; and linear Fresnel reflector systems, where one receiver tube is positioned above several mirrors allowing the mirrors a greater mobility in tracking the sun.

Parabolic through system **Linear Fresnel reflector system**

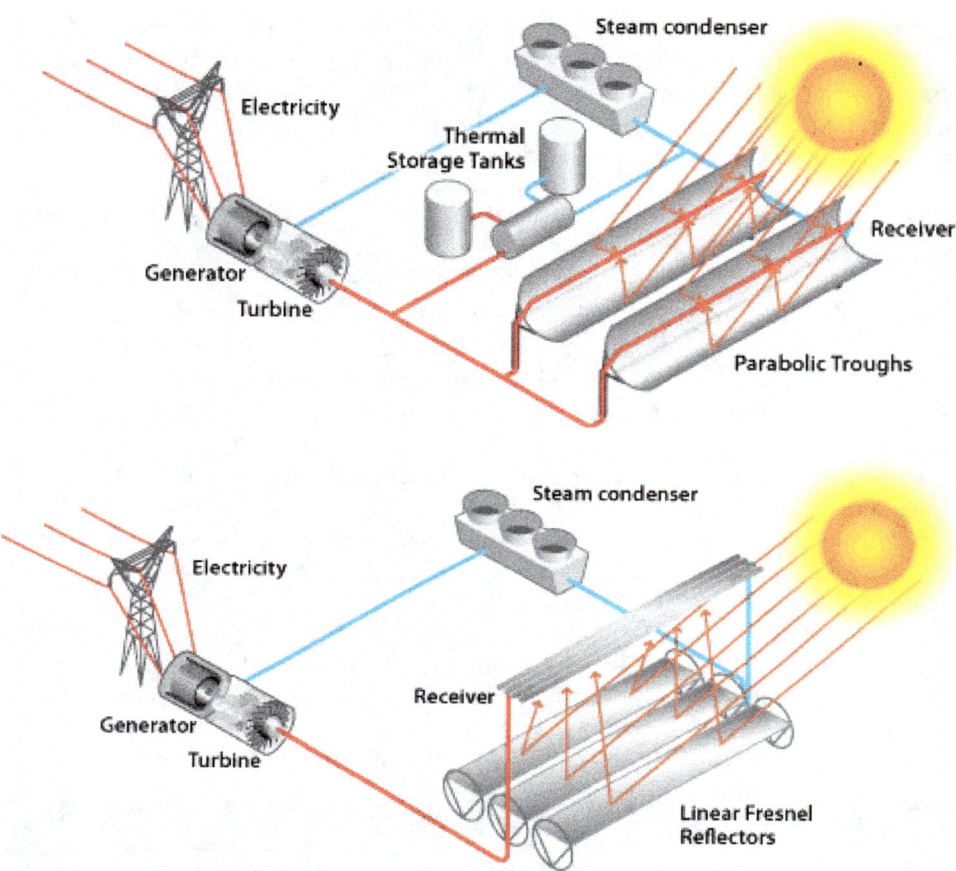

Dish/Engine systems use a mirrored dish like a large satellite dish. The mirrored dish is usually composed of many smaller flat mirrors formed into a dish shape to minimize cost. The dish-shaped surface directs and concentrates sunlight onto a thermal receiver, which absorbs and collects the heat and transfers it to the engine generator. The most common type of heat engine used in dish/engine systems is the Stirling engine. This system uses the fluid heated by the receiver to move pistons and create mechanical power. The mechanical power is then used to run a generator or alternator to produce electricity.

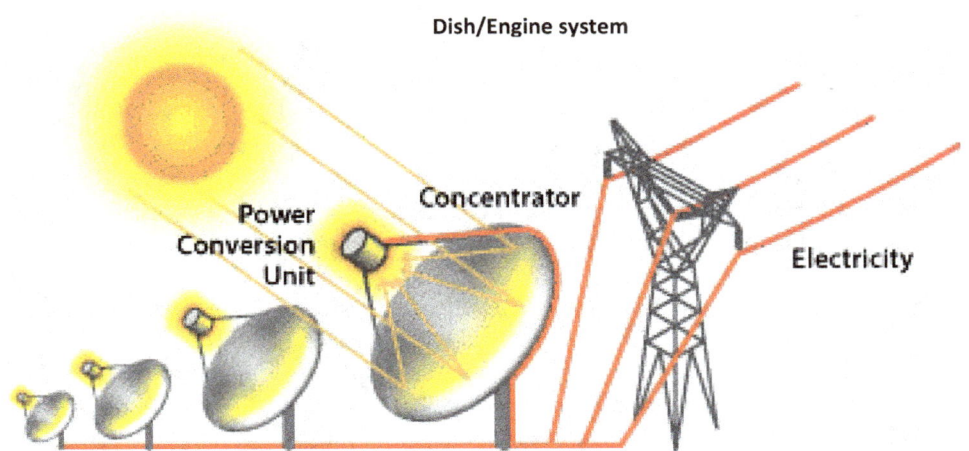

Dish/Engine system

Power Tower systems: A power tower system uses a large field of flat, sun-tracking mirrors known as heliostats to focus and concentrate sunlight onto a receiver on the top of a tower. A heat-transfer fluid heated in the receiver is used to generate steam, which in turn, is used in a conventional turbine generator to produce electricity. Some power towers use water/steam as the heat-transfer fluid. Other advanced designs are experimenting with molten nitrate salt because of its superior heat transfer and energy storage capabilities. The energy storage ability or thermal storage allows the system to continue to dispatch electricity during cloudy weather or at night.

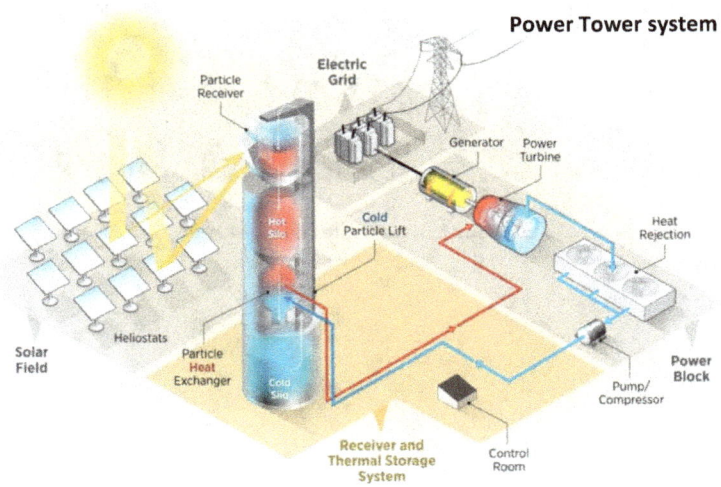

Power Tower system

The Ivanpah Solar Electric Generating System, in California's Mojave desert, is the largest concentrated solar thermal plant in the world. It was constructed over four years from 2010 to 2014 at USD 2.2 billion. The plant produces about 392 MW of electricity using 173,500 software-controlled mirrors, called heliostats spread over an area of 5.5 miles². Each heliostat has two mirrors that follow the sun and reflect it onto water-filled boilers atop three separate 450-foot towers. When the sunlight hits the boilers, the water inside is heated and creates high-temperature steam. The steam is then piped to conventional steam turbines, which generate electricity.

Aside from the U.S., Spain has several power tower systems. Planta Solar 10 and Planta Solar 20 are water/steam systems with capacities of 11 MW and 20 MW, respectively. Gemasolar, previously known as Solar Tres, produces nearly 20 MW of electricity and uses molten-salt thermal storage.

Ivanpah Solar Electric Generating System, California's Mojave desert

Geothermal Energy

The word "Geothermal" comes from the Greek word geo, meaning earth and thermal, meaning heat. Thus, geothermal means heat energy within the earth that exists in reservoirs of hot water and steam deep within the sub-surface of the earth. Geothermal energy originates from the original formation of the planet and the slow decay of radioactive particles in the earth's core. It is believed that the earth's internal temperature at the core can reach up to 10,000 ºF (5,537 ºC), which is as hot as the sun's surface. These extreme temperatures melt any solid material, such as iron and rocks, into a glue-like viscous fluid. Since the resulting fluid is lighter, it rises upwards towards the earth's crust through fractured or faulty tectonic plates. This is the source of geothermal energy that has been exploited for generations.

Archaeological evidence shows geothermal energy usage by all ancient cultures around the world. For example, there is evidence of Native Americans using hot springs for heating, cleaning, and healing as far back as 10,000 years ago. The recorded history of geothermal energy use, however, goes back only 2000 years. The Chinese, Romans, and Turkish are among the many cultures where hot springs were used for bathing and heating. In modern times, geothermal energy was used for district heating in the late 1400s in France followed by a similar district heating system in America in Boise, ID in the 1890s.

The early 1900s saw the rise of geothermal energy which was used for heating buildings and greenhouses. Hot Lake Hotel, in Union County, Oregon, became the first known building in the world to use geothermal energy to heat the building in 1907. This was followed by geothermal heating of greenhouses in Boise, Iceland, and Italy in the 1920s. The early 1940s witnessed the use of steam and hot water from geysers for heating residential homes in Iceland. Today, Iceland is the world leader in this practice, with over 90% of its homes heated with geothermal energy, saving the country over $100 million annually in oil imports. In fact, Reykjavík,

Iceland has the world's biggest district heating system, often used to heat pathways and roads to hinder ice accumulation.

The early 1900s also witnessed the rise of geothermal energy for electricity production. The first known geothermal power plant was tested in Italy in 1911, followed by the first industrial geothermal electricity production in 1958 in New Zealand. Today technological advancement has allowed many nations to add geothermal energy to their electric grid system. Unfortunately, geothermal energy remains under-utilized due to a lack of investment and financing. For example, in the United States, 3.6GW of geothermal electricity (enough to power nearly 3 million households) was produced in 2022 accounting for only 0.4% of total electricity production. Nearly 95% of the 3.6GW of electricity produced came from two states; California and Nevada. The remaining 5% came from an additional five states. All seven states that account for 100% of U.S. geothermal energy are in the western region. Obviously, their tectonic nature gives them an advantage over the other states. Globally, a handful of countries accounted for nearly 75% of global geothermal electricity produced in 2022. These countries are the U.S., Indonesia, Philippines, New Zealand, Mexico, and Turkey.

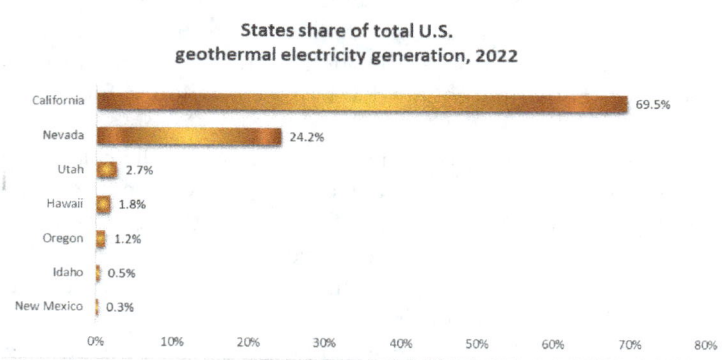

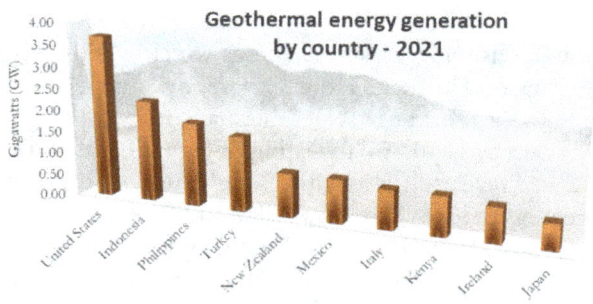

Source of geothermal energy

Like the energy from the sun, heat from earth's interior is infinite. Thus, geothermal energy is an infinite and renewable source of energy. To understand why geothermal energy is considered renewable, we need a basic understanding of the earth's interior

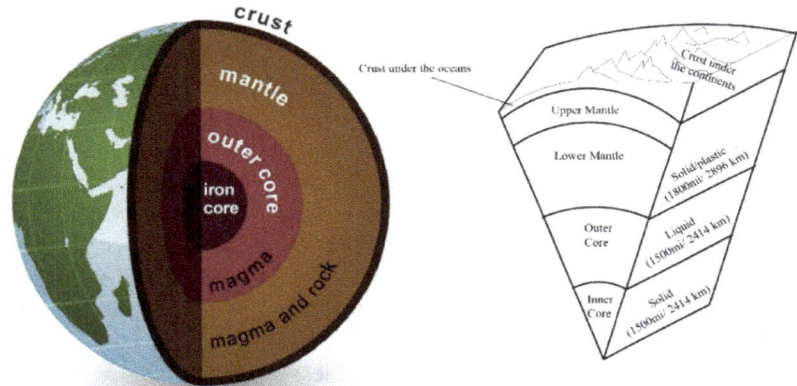

Source: National Energy Education Development Project graphic

formation. The earth has four major layers comprised of different materials, such as iron, nickel, and silicate. These layers are inner core, outer core, mantle, and crust. The extreme temperature and pressure that occurs in the earth's interior affects the melting point of the materials and results in the different characteristics of the layers.

An inner core is the deepest layer made up of solid iron/nickel that is about 1,500 miles (2414km) thick. The temperature of the earth's inner core ranges from 7,000 ⁰F to 11,000 ⁰F, which is as hot as the surface of the sun. The intense pressure that exists in the inner core layer changes the melting point of the iron/nickel, allowing it to remain solid, despite the extreme temperature.

An outer core is an outer layer of the core made up of hot molten iron/ nickel called magma. It is about 1,500 miles (2414km) thick. This layer is liquid due to the high temperatures and the decreased pressure compared to the inner core.

A mantle is a layer of magma and rock surrounding the outer core. It is about 1,800 miles (2890km) thick making it the thickest layer. Temperatures in the mantle range from about 392 °F at the upper mantle-crust boundary to approximately 7,230 °F at the mantle-outer core boundary. The mantle layer contains solid/glue-like silicate material. The upper mantle is solid silicate due to lower temperature and pressure, while the lower mantle is molten and has lower viscosity due to increased temperature. The heat from this layer escapes and surfaces up towards the crust through fractured or faulty tectonic plates heating ground water in the pores and fractures of the rocks. Groundwater, which is rainwater that has penetrated deep into the fractured crust, is heated by the hot rocks into a large, heated water and steam reservoir.

A small portion of the hot water and steam mixture rises to the surface through the fractured rocks creating hot geysers and hot springs. The pathways or the fractured system channels that the hot water and steam mixture travels through determine the speed and the chemical composition of the mixture. A straighter pathway to the surface leads to fast-shooting waters that have retained their original chemical composition of pure water. The geysers of Yellowstone National Park in the United States are good examples.

However, a less straight pathway to the surface leads to a hot water/steam mixture that rises upwards at a slower flow rate. When rising upwards at a slower rate, the mixture is more likely to mix with the chemical composition of the variety of rocks beneath the surface. Thus, the mixture that reaches the surface is comprised of water and minerals from the rocks. This is witnessed by the many colorful rocks and acid lakes in Yellowstone National Park and other parts of the world, such as the Danakil region of Coastal Eritrea.

Acid lake, Yellowstone National

Colorful rocks and lakes, Danakil, Eritrea

A crust is the outermost layer made up of solid rock that forms the continents and ocean floors. The layer is 15 to 35 miles (24km-56km) thick under the continents and 3 to 5 miles (5km-8km) thick under the oceans. As previously mentioned, silicate materials make up over 90% of the earth's crust, while nickel/iron alloys make up the rest. The low temperature and pressure that exists at this layer allow the crust to be solid.

The earth's crust is broken into pieces called tectonic plates found under the continents and ocean floors. The plates drift apart and push against each other at the rate of about one inch per year in a process called continental drift. Continental drift creates faults and fractures in the crust. Magma comes close to the earth's surface near the edges of these plates through the faulted or fractured routes. The lava that erupts from volcanoes is partly magma.

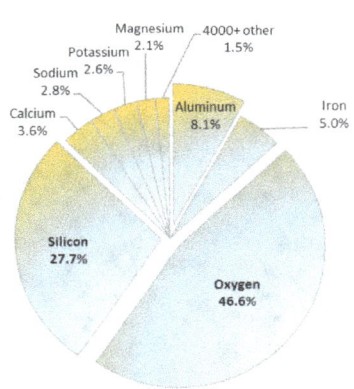

Exploration

Exploration refers to the process of gathering information regarding a potential geothermal resource area. Since magma flows upwards to near the edge of the tectonic plates, it makes sense to look for geothermal energy along these lines. Scientists have identified seven major tectonic plates along with many more minor and micro plates. The seven major plates are North American, South American, Pacific, African, Antarctic, Eurasian, and Indo-Australian. Most large-scale geologic events, such as volcanoes or earthquakes, occur along these tectonic plates which exist along the edges of the continents, island chains and beneath the sea. In fact, more than half of the world's active volcanoes above sea level occur along the island chains that encircle the Pacific Ocean to form the circum-Pacific "Ring of Fire".

Most of the previously discovered and/or untapped geothermal energy exists along the lines of the tectonic plates. For example, two of the major moving tectonic plates (the North American and the Pacific plates) meet on the coast of California. The boundary where they meet is a major earthquake zone, known as the San Andreas fault. The entire San Andreas fault system extends over 800 miles long and at least 10 miles deep inside the earth's interior. This area has provided the state of

California with tremendous potential for geothermal energy. In fact, the largest geothermal energy plant in the world, the Geysers geothermal complex, is in this area. The plant generated between 700MW to 900MW, enough to power nearly 1 million homes. California accounted for nearly 70% of U.S. geothermal energy in 2022 The state of Nevada accounted for an additional 25% of the total U.S. geothermal electricity output that same year.

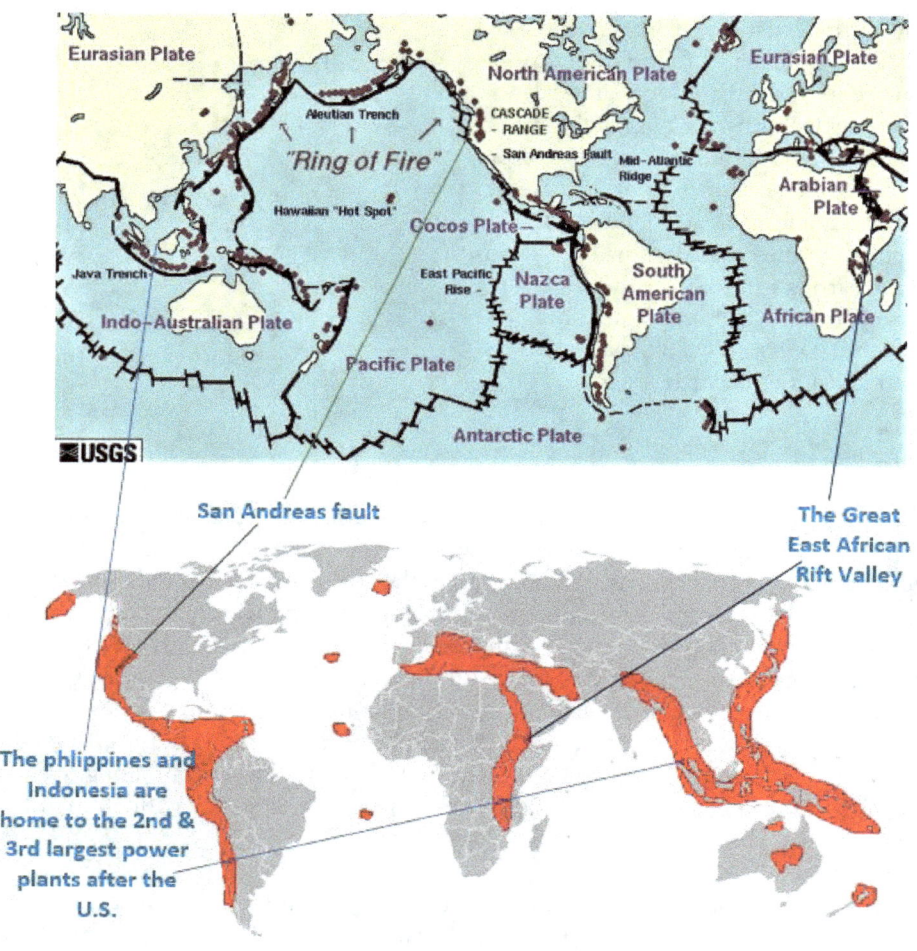

(Top map) Untapped geothermal energy exists along the lines of the tectonic plates
(Lower map) the most developed geothermal plants exist along the red highlighted areas

The map below shows favorable areas for geothermal energy development in the United States.

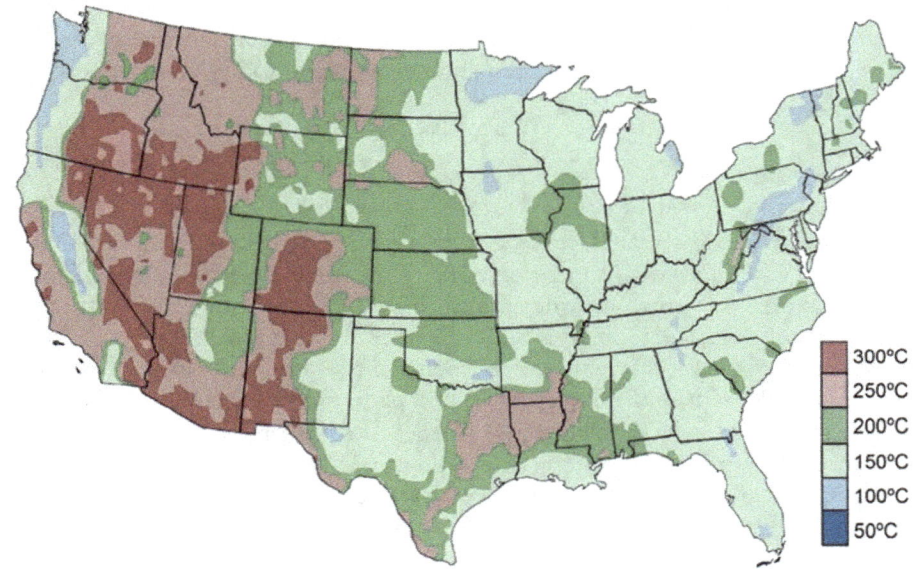

U.S. Geothermal Resources at 10 KM depth.
States in the west have greater potential for geothermal energy growth.

All regions along the edges of the tectonic plates offer untapped geothermal energy. The East African Rift System (EARS), where the African plate and the Arabian plate meet is one of those promising areas. The rift extends thousands of miles from Lebanon in Asia, past the Red Sea before turning inland into the Danakil region of Eritrea and Ethiopia, making the region one of the most active volcanic zones on earth. The rift continues towards the Ethiopian highlands into Kenya, Uganda, the southern tip of South Sudan, and Rwanda. Finally, the rift runs along two separate branches (Western and Eastern Rift Valley), touching Tanzania, Zambia, and Mozambique.

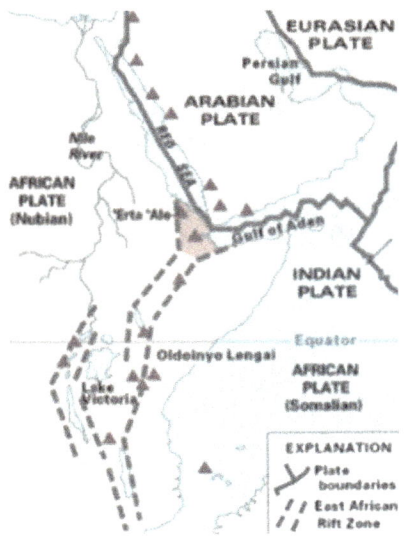

Kenya is the only nation that has invested in the geothermal resources that exist in the region. Currently, Kenya is ranked as the 8th largest geothermal-producing country in the world. The country generates nearly 1 GW of electricity, which meets 46% of its electricity demand. Other neighboring nations, such as Eritrea and Uganda, are heavily involved in research and development to tap into this abundant energy source.

Exploration Methods

Once a potential geothermal resource field is identified, several methods are used to explore and gather detailed information. One of the methods is called *geophysical*. Geophysical methods provide sufficient information about the subsurface structures of geothermal fields. The method is used to gather information about the physical characteristics of the geothermal reservoir beneath the surface, including the depth of the reservoir, fractures that control the fluid path, fluid temperature and pressure, as well as rock porosity. The electrical properties of rocks provide important information like temperature, geothermal fluid characteristics, porosity, and whether the rocks are weathered or un-weathered which directly impact the geothermal systems.

Geophysical method applies seismic, electrical resistivity, potential (gravity and magnetic), temperature gradient and heat flow measurements. The most common methods are electrical resistivity and electromagnetic.

Electrical resistivity method relies on a signal difference between two points. Direct electrical current (DC) is passed into the earth through electrodes. This produces an electrical field which is measured as the potential difference between the electrodes. The electrodes by which current is introduced into the ground are called Current electrodes and electrodes between which the potential difference is measured are called Potential electrodes. The electrical field generated is measured at the surface and evaluated based on Ohm's law:

$$\text{Electrical field strength (E)} = pj \text{ (V/m)}$$

j=current density (A/m^2)

p=electrical resistivity is a measure of resistance or the power of certain material, such as rocks, to resist the flow of moving current. Resistance is defined as voltage divided by current ($R=V/I$).

In the above example, the setup for a resistivity survey is performed using a resistivity meter and four electrodes. The resistivity meter is a device that acts as both a voltmeter (V) and an ammeter (I) to measure current and records resistance values (V/I). The current (AB) and the potential electrodes (MN) are equidistant from the center. The potential electrodes are kept close to the center at a fixed distance while the current electrodes are positioned away from the center and moved away from the center after each measurement. The greater the distance between the current electrodes, the deeper the current will flow. Finally, a cross-

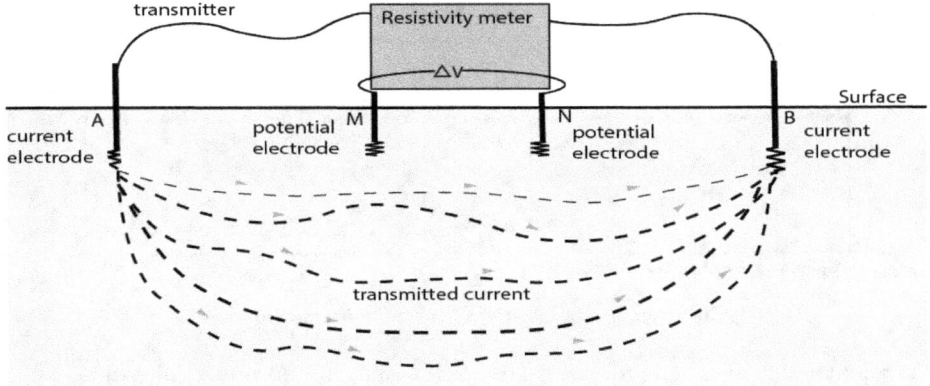

section of resistivity levels with increasing depth is graphed and compared to the resistivity levels of different materials. The cross-section allows for determining the physical characteristics of what is beneath the surface.

Rock Material	Resistivity (Ohm-meter)
Top soil	50 - 1,000
Granite	300 - 20,000
Basalt	200 - 20,000
Marble	100 - 200,000
Quartzite	10 - 200,000
Shale	80 - 20,000
Sandstone	30 - 500,000
Clay-shale	0.00004 - 900
Limestone	30 - 500,000
Clay	1 - 100
Fresh groundwater	10 - 100
Loams	10 - 450
Sand	100 - 500,000
Oil sands	60 - 900

List of sample materials and their resistivity.

Electromagnetic methods are used to detect variations in subsurface electrical resistivity and determine the electrical conductivity and hydrocarbon content of materials beneath the earth up to a few kilometers in depth. This method is especially useful in the search for hydrocarbons as in the oil and gas industry. In the geothermal energy industry, this method is helpful in searching for steam or water at a depth of up to 4km where it is considered suitable for heating of buildings, greenhouses, or other projects. Electromagnetic method is accomplished in two different ways: transient electromagnetic (TEM) and magneto telluric (MT). TEM, also known as time-domain electromagnetics (TDEM), works by passing an electric current into the earth through a time-varying magnetic field from a controlled source. Basically, the current-producing device transmits current into its transmitter coil, or loop. As the current travels in the transmitter loop, it generates a magnetic field that has the same frequency and phase as the current. This induced field propagates lines of force that penetrate the earth. After the transmitted signal is turned off, the decaying magnetic field or the voltage returned from the earth's materials is measured at the surface as a secondary magnetic field. The returned secondary energy is then compared to the initial transmitted energy. The data is then processed and analyzed to learn the electromagnetic (EM) characteristics of the material beneath the surface.

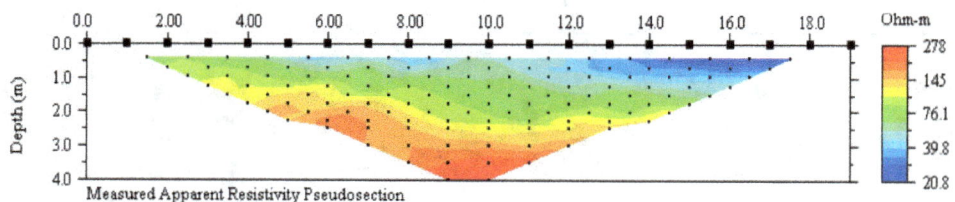

Cross-section of resistivity level with increasing depth

Transient electromagnetic methods are noninvasive and do not require direct electrical contact with the ground. Thus, it is ideal for moving or aerial surveys as well as for use at sites with very resistive surface layers. Depths of investigation (DOI) vary from tens to hundreds of meters below the land surface.
MT method is similar to TEM but allows the detection of resistivity at a much deeper depth ranging from 100 meters to 200 km. The method is used to detect the electrical conductivity/resistivity associated with geothermal structures, such as faults and reservoir temperatures. MT is one of the most common methods used in exploration surveys around the world. It is used for geothermal, oil and gas, and mining explorations as well as for groundwater monitoring. MT method has been used to identify geothermal energy in many parts of the world, including Japan, the Philippines, the United States, Iceland, and China.

Extraction

Since heat is continuously produced inside the earth, geothermal energy is renewable energy. Geothermal energy is extracted from the earth by pumping fluids into the heated rocks or heated reservoirs that occur naturally at some depth below the earth's surface. The heated fluid is then pumped back to the surface and used as steam to run a generator. The extraction process is accomplished in two different ways: Hydrothermal systems and Enhanced Geothermal Systems (EGS).

Hydrothermal system is naturally occurring where water is used as the circulating fluid. Groundwater or rainwater penetrates deeper into the crust through rock fractures and faults. The water is heated by the hot rocks. The hot water/steam mixture then rises to the surface as hot springs. Hydrothermal energy sources are site-specific, and their occurrences are controlled by geological and tectonic configurations. Thus, this system applies to only a few places around the world where there is a rich reservoir of thermal energy. Locations suited for hydrothermal systems are classified into six groups.

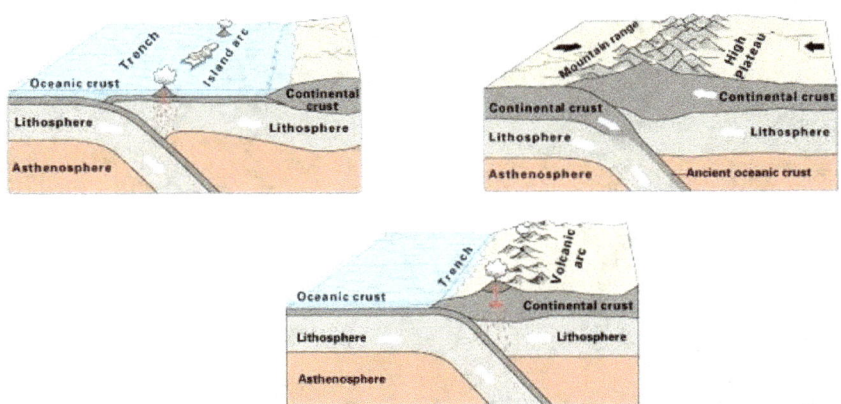

a) subduction-related volcanic settings: Ocean vs ocean plate, ocean vs continental plate. Indonesia and The Philippines are examples of where this kind of volcanic activity is present. For example, Sumatra, Indonesia hosts nearly 200 volcanoes with many of them still active. Indonesia has an estimated 27 GW of potential geothermal energy and only 4% is being exploited.

b) continental collision zones: Himalayan geothermal province with nearly over 100 thermal springs and steam vents in an area of 1500km. The presence of granitic melts at shallower levels and shallow mantle depth and the frictional heat resulting from thrust are the main sources of heat for geothermal systems.

c) infra-continental rifts and ocean rifts volcanic settings, such as the East African rift valley (Eritrea, Ethiopia, Kenya, Djibouti, and Saudi Arabia)

d) infra-continental rifts without volcanism
   Larder Ello, Italy and the West coast of India. Larder Ello is one of the few places on earth that produces superheated steam due to mass transfer of heat from the rocks to the fluid. Italy produces 914MW of which 100 MW is in larder Ello.

e) geo-pressured systems associated with oil fields

f) crust-mantle boundary disequilibrium settings

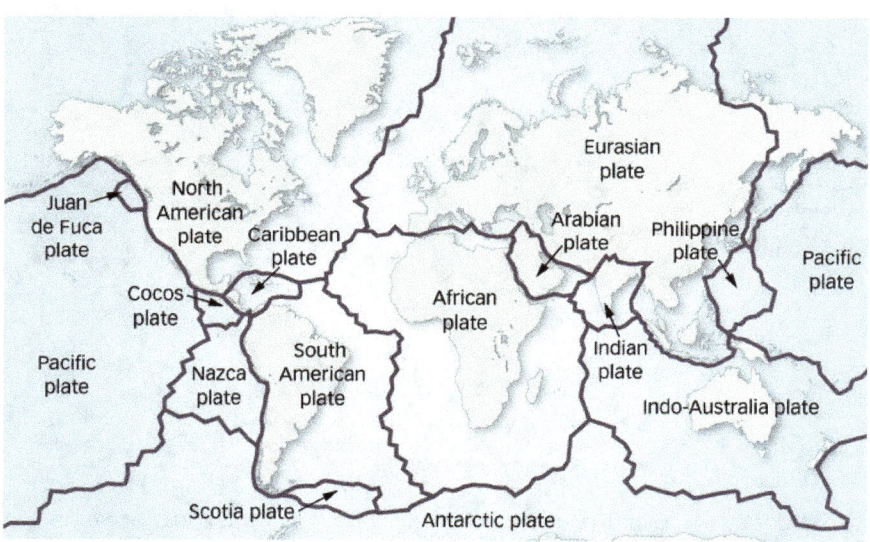

hydrothermal systems occur along the lines of tectonic plates

Enhanced Geothermal System (EGS) also known as man-made geothermal system or engineered geothermal system, is a technology that has been developed since the 1970s. The system works by drilling two holes in the deeper crystalline rocks and connecting them by a network of fractures. Then pressurized fluid, mainly water is circulated through the drilled holes. The initial task involves identifying an area where the rocks, at a reasonable depth, are sufficiently hot, with low permeability and can support the construction of a loop to circulate water through the induced fractures. Granite rocks are excellent candidates to meet the requirement. The decaying process of the naturally occurring radioactive elements, such as uranium, thorium, and potassium, that exist in the rocks generates the desired heat. Finally, water is injected under high pressure to expand existing rock fissures to enable the heated fluid or steam to flow in and out freely.

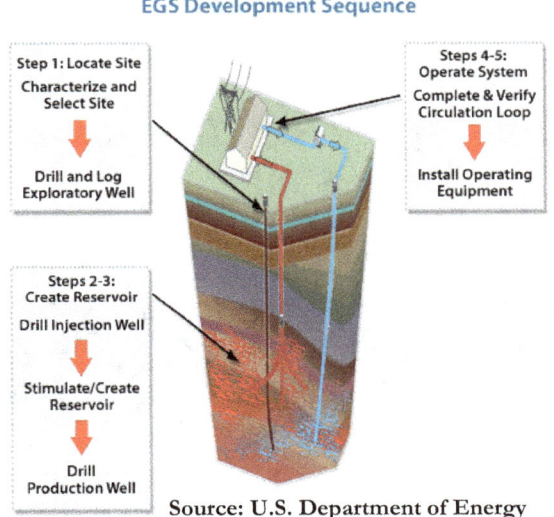

Source: U.S. Department of Energy

Since radioactive elements are naturally occurring at shallower depths of 3km to 5km in the crust, thermal energy can be produced at shallow depths using EGS nearly anywhere in the world. The technique, which is adapted from the oil and gas industry, allows for wells as deep as 6 miles (10km) compared to the 2 miles (3km) of depth allowed during previous methods. The heat from the radioactive elements, in addition to the heat conducted by the earth's interior, provides an excellent energy source that can be extracted. With the help of directional drilling, a large area of granite below the surface can be reached.

EGS technology is now mature and is commonly used worldwide to get unlimited supply of clean and carbon-free electricity. EGS, like hydrothermal energy sources, can provide continuous baseload power with minimum impact on the environment. Geothermal energy does not pollute the air and no byproduct or waste is generated. CO_2 emissions are negligible and may even be used as the working fluid to extract heat from the granite reservoir. Furthermore, the land requirement for both hydrothermal and EGS, unlike other renewables, is very small.

Geothermal Power

Harnessing geothermal energy to generate electricity (Power generating) is a relatively new and rapidly growing technology. The process starts with the completion of geological and geophysical surveys. Following a successful survey, a production well is dug up and allowed to bring steam or thermal water to the surface. Then the endurance of the production well and the reservoir is tested by allowing steam or water to eject to the surface over a long period. This allows enough time for the evaluation of changes in the flow rate, temperature, or pressure.

Once the well has passed the test, it goes to the power production state where it is ready to generate electricity. There are three different kinds of technologies used in geothermal power plants today: Dry steam power plant, Flash steam power plant, and Binary cycle power plant.

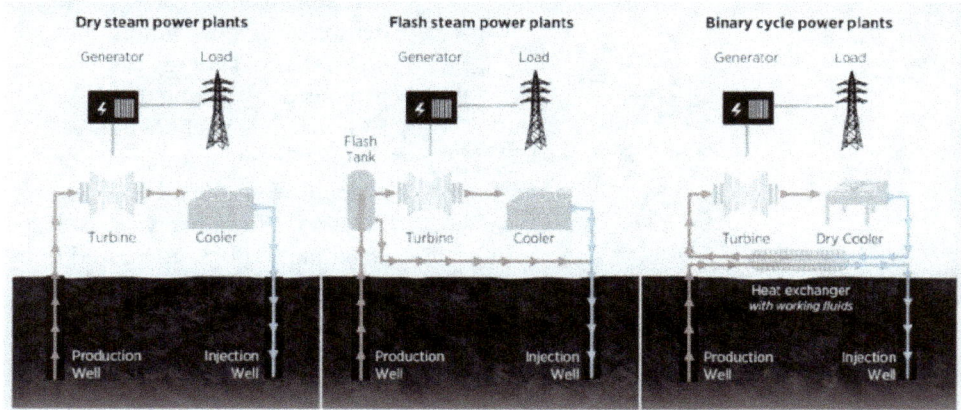

Geothermal fields either produce only steam, water, or a mixture of both. If the geothermal system supplies only steam, then a dry steam power plant design is adopted for running the turbine. Steam from each well is gathered using a network of insulated pipes to a central receiving station and is directed to a turbine. The steam pipes are often bent in a loop to allow for thermal expansion. Some pipes are over 5km long and expand to about 5m. If the well produces two phases (steam and water), both phases are utilized in flash steam power plants for generating electricity. If the geothermal field produces only thermal water, then binary cycle technology is commonly adopted. Any single technology or a combination of all three may be used.

Dry steam power plan involves using steam drawn directly from high-temperature reservoirs deep beneath the earth. Steam is then used to spin a turbine which drives an electricity-producing generator. After the steam condenses, it is cooled and reinjected back into the reservoir to be heated up again and pumped up to the surface as

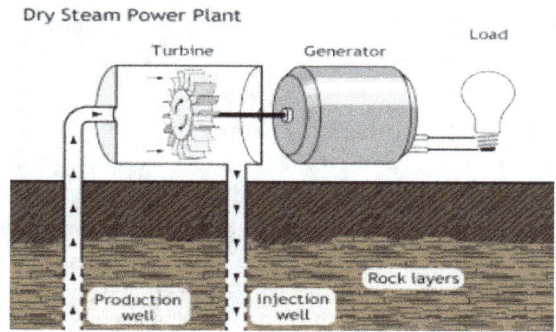

steam. This technology is the simplest form of geothermal power plant and has been used since 1904. However, since naturally occurring stream reservoirs are rare, dry steam power plants are very few around the world. The largest geothermal power facility, The Geysers geothermal plant complex in California is the only dry steam facility in the United States. Another dry steam source is in Yellowstone National Park. However, since this is a protected national park, it is used for tourism and not for commercial electricity production.

Flash steam power plants use extracted hot water or a mixture of hot water and steam (greater than 360oF/182oC). The steam is directly routed towards the turbine and generator for electricity production while the hot water is pumped under high pressure into a tank at the surface. The tank is held at a much lower pressure, causing some of the fluid to rapidly vaporize or "flash." The vapor then drives a turbine, which drives a generator. If any liquid

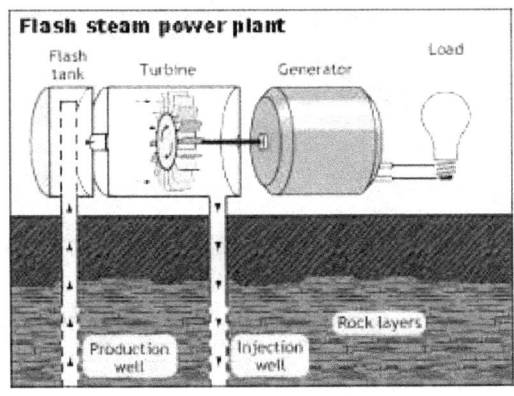

remains in the tank, it can be flashed again in a second tank to extract even more steam/energy. This is called a double flash instead of single a flash power plant.

In the case of a single flash system, the steam or steam and water mixture from the production well is directed to a cyclonic pressure vessel where the two phases with distinct density contrast get separated. The separated steam is then directed to run the turbine while the water is injected back into the aquifer through an injection well. If there are several production wells, then the cyclone separator can be placed at each well or near the powerhouse, where all the stem and water are collected by a network of pipes. A single well can generate about 4 to 6 Mwe. After the steam separation in a steam separator, the separated liquid may still contain steam with sufficient enthalpy. In such cases, the hot water is flashed to extract additional steam in a flasher and steam thus separated is directed to a low-pressure turbine to generate electricity. The power plants of this type are called double flash power plants.

The steam generated from the flasher is of low pressure and is therefore used for generating power from a low-pressure generator. Currently, the turbines are designed to receive both low and high-pressure steam (dual admissible turbine) and thus, reduce the capital cost of the power project. The advantage of a double flash steam plant is that it can extract more steam from the system and can generate additional power of the order of 15 to 25% (Di Pippo, 2008). Flash steam plants are the most common type of geothermal power generation plants in operation today due to their flexibility in using low-temperature hot water as a source.

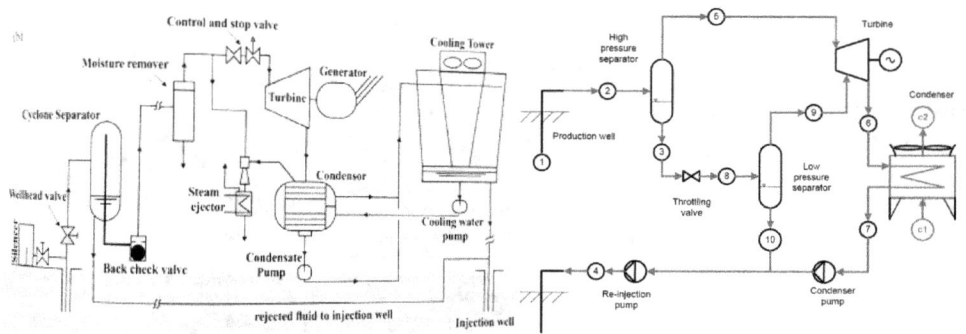

(Left) single flash geothermal power plant (Right) double flash plant
Source: Di Pippo, 2008

Binary cycle power plant is utilized in conditions where the hot water is not fit for flashing. In this case, the heat from the hot water is extracted through a heat exchanger to generate power using binary cycle technology. Binary cycle power plants were developed in the Soviet Union in the 1960s and allowed for electricity to be generated from much lower-temperature water than the previously used heat pumps of the 1940s. Hot water from production wells as low as 76 ^{0}C can be used to generate electricity. For this reason, a significant proportion of geothermal electricity in the future could come from binary-cycle plants.

In a binary cycle power plant (also known as Organic Rankine Cycle or ORC), water or other geothermal fluid is actively injected into wells to be heated and pumped back out. The heated fluid passes through a heat exchanger in a Rankine cycle binary plant and vaporizes another organic fluid known as the working fluid, which has a much lower boiling point than the geothermal fluid. In principle, heat is extracted from the geothermal fluid through a heat exchanger and transferred to the working fluid. The newly vaporized working fluid is then used to drive a turbine that spins an electricity-producing generator.

The general criteria for the selection of a working fluid are that the fluid should be non-corrosive, non-flammable, and should not react at the temperatures and pressures it is exposed to. The organic fluids commonly used in the ORC technology are n-butane (BP: -0.5⁰C), i-butane (BP: -11⁰C) for thermal fluids with temperatures <200⁰C, and toluene (BP: 1100⁰C) for thermal fluid >200⁰C. (additional search).

These organic fluids are, in fact, greenhouse gases and hence have the potential to cause environmental degradation. However, since Binary cycle power plants are closed-loop systems, small amounts of toxic gases (except water vapor) are emitted into the atmosphere. In addition, there have been technological advancements, such as the Kalina cycle, to help minimize any environmental degradation caused by organic fluids.

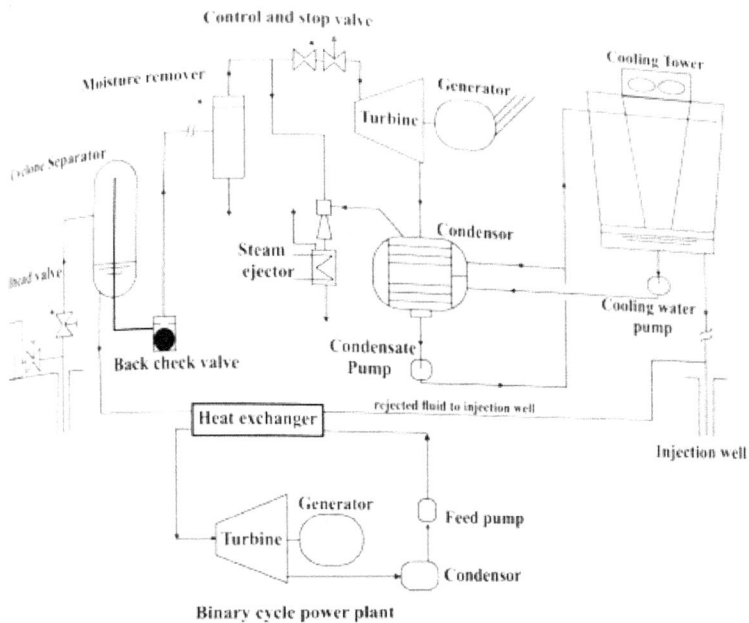

**A combined single flash and binary cycle geothermal power plant.
Source: Di Pippo, 2008**

Kalina cycle utilizes a mixture of Ammonia, which is not a greenhouse gas, and water instead of the previously mentioned organic fluids. The technology was invented by a Russian scientist Dr Alexander Kalina in 1983 and used mainly to run engines. The design was later modified to suit several industries, including the geothermal industry. Since ammonia and water have different boiling temperatures with near similar molecular weights (ammonia: 17kg/kmol, water 18kg/kmol), both liquids can be separated with ease. The ammonia-water mixture evaporates over a wide range of temperatures, depending on the mixture ratio. This binary fluid is best suited for generating power from high and low-enthalpy geothermal sources.

In addition, a mixture of ammonia and water has a higher volatility, and thus a higher efficiency compared to the ORC plants. Increased heat transfer from geothermal fluids to the mixture leads to slightly higher power output. The following charts are an analysis of output power, efficiency, and heat transfer rate when organic fluid (ORC) is used versus ammonia/water mixture as a working fluid. The charts are for geothermal fluids above 110°C.

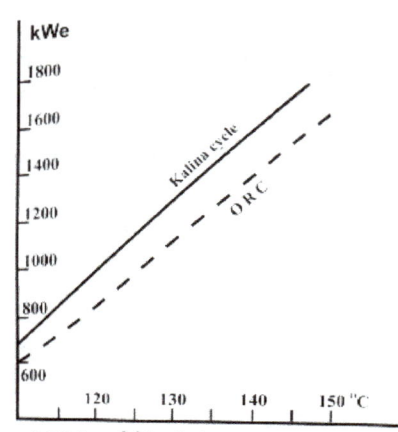

Inlet water temperature vs power output (kWe)
Source: Chandraseiver and Bundschuh, 2008

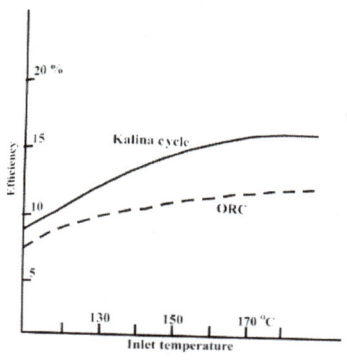
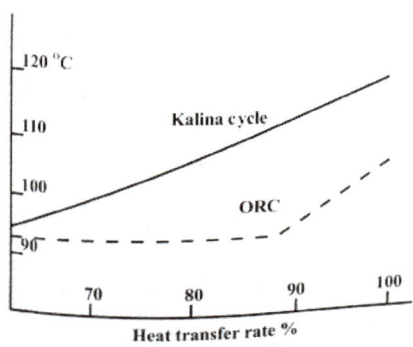

(Left) inlet temperature vs Efficiency. Source: Chandraseiver and Bundschuh, 2008
(Right) Heat transfer rate (%) vs temperature. source: Kalina, 1984

Direct Applications

Technologies for direct geothermal consumption are widely used and considered mature. As previously mentioned, the concept has been used for centuries for many purposes, including space cooling and heating for individual homes, district heating, dehydration and greenhouse cultivation, aquaculture, industrial processing, and for balneology (therapeutic bathing). Thermal energy from underground reservoirs where heat temperatures are between 80^0F- 300^0F (26^0C–148 °C), is directly used for heating in homes, industries, greenhouses, and for bathing in many places. In addition, the thermal energy from shallow reservoirs, where temperatures are between 50^0F -70^0F (10°C to 20°C), is extracted with heat pumps for similar purposes of space heating.

Ground source heat pump (GSHPs)

The more commonly used direct geothermal is space heating and cooling using heat pumps. Because of its long history of application, geothermal heating is cost-effective and thus, has increasingly become the preferred energy choice for home heating. One of the more commonly used methods of heating and cooling homes is Ground Source Heat Pump (GSHPs). GSHPs transfer heat to and from the earth to provide cooling and heating for homes and buildings. During winter, heat is pumped up into the house while colder airflow is pumped down to the ground. The reverse takes place during summer. The process requires digging wells underground where the temperature fluctuation is minimal. Thus, the earth is a source and a sink of heat. The medium of the heat carrier or heat remover is a liquid, either groundwater or a refrigerant aided by a heat pump. During the summer months, heat is removed from the homes and transferred to the earth while in the winter months, heat is transferred to the homes. The system takes advantage of the earth's inherent thermal behavior – in summer the ground is cooler than the surface temperature and during winter it is warmer.

GSHPs are accomplished in closed-loop or open-loop systems.

Closed-loop systems use antifreeze solution as a circulating fluid in high-density plastic-type tubing. The tubing is buried in the ground or submerged in water. A heat exchanger transfers heat between the refrigerant in the heat pump and the antifreeze solution in the closed loop. The heat exchanger, a loop of coil, can be placed horizontally or vertically below the ground depending on the availability of space, soil condition, climate, and cost of installation.

Vertical loop has the advantage of providing a constant mean earth's temperature controlled by a geothermal gradient. It also has the advantage of being installed in a space-constrained environment. Thus, it is more suitable for large commercial buildings and schools where land space is limited. Vertical loops are also used where the soil is too shallow for trenching. Installation consists of holes (approximately four inches in diameter) drilled about 20 feet apart and 100 to 400 feet deep. Two pipes, connected at the bottom with a U-bend to form a loop, are inserted into the hole and grouted to improve performance. The vertical loops are connected with horizontal pipes placed in trenches and connected to the heat pump in the building.

Horizontal loop, on the other hand, has the advantage of being placed at shallow levels. This installation is most cost-effective for residential installations, especially for new construction where land space is available. It requires trenches at least four feet deep. The most common layouts either use two pipes, one buried at six feet, and the other at four feet or two pipes placed side-by-side at five feet in the ground in a two feet wide trench. The Slinky method of looping pipe allows more pipe in a shorter trench, which cuts down on installation costs. A disadvantage of a horizontal loop, however, is that it is affected by the surface temperature of the earth which is affected by rain, solar radiation, and wind.

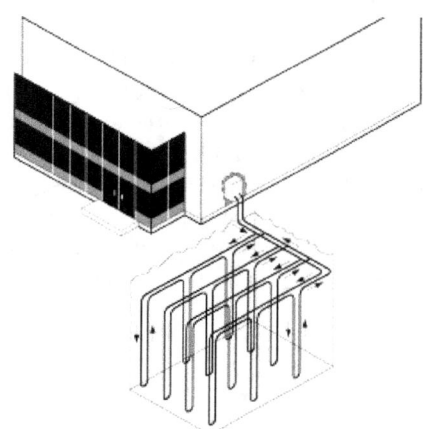
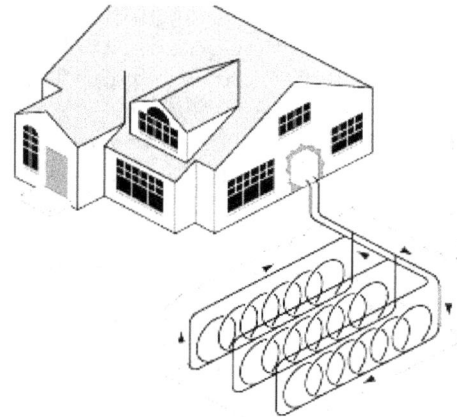

A diagram of vertical closed loop system.
Source: U.S. Department of Energy

A diagram of horizontal closed loop system.
Source: U.S. Department of Energy

In addition, closed-loop systems can be installed in a **pond/lake** option where there is an adequate body of water. A supply line pipe is run underground from the building to the water and coiled into circles at least eight feet under the surface to prevent freezing. This is one of the lowest installation types of GSHP. **Direct exchange** is also another closed-loop installation where a heat exchanger is not used. Instead, the refrigerant is pumped through copper tubing that is buried in the ground in a horizontal or vertical configuration. These installation systems work best in moist soils.

Open-loop system uses groundwater or surface body water as the main circulating fluid through the GSHP system. Once the fluid has circulated through the system, the water returns to the ground. Thus, the system is limited by the geological condition of the project area. There must be an adequate supply of relatively clean water. In addition, local codes and regulations regarding groundwater must be considered.

Another technology used as an open-loop system is **Hybrid systems**. This method uses several different geothermal resources, or a combination of a geothermal resource with outdoor air, such as a cooling tower. Hybrid approaches are particularly effective where cooling needs are significantly larger than heating needs.

Where local geology permits, the "standing column well" is another option. In this variation of an open-loop system, one or more deep vertical wells are drilled. Water is drawn from the bottom of a standing column and returned to the top. During periods of peak heating and cooling, the system can bleed a portion of the return water rather than reinjecting it all, causing water inflow to the column from the surrounding aquifer. The bleed cycle cools the column during heat rejection, heats it during heat extraction, and reduces the required bore depth.

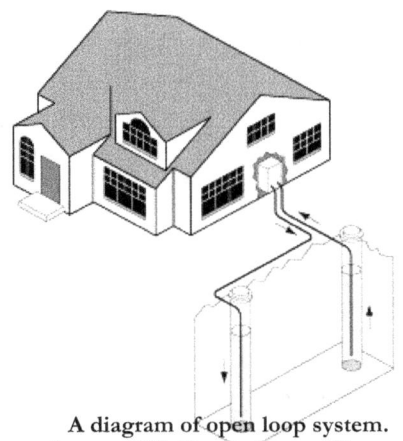
A diagram of open loop system.
Source: U.S. Department of Energy

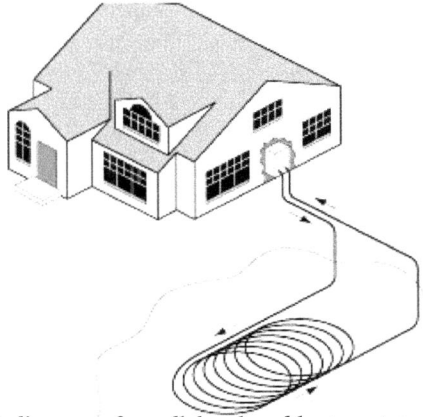
A diagram of pond/lake closed loop system.
Source: U.S. Department of Energy

All of the above-mentioned approaches can be used for residential and commercial building applications. Environmentally, the impact of the heat pumps is minimal. The system needs a small input of electricity. If the electricity supply is from a renewable source, then the net output of CO_2 is negligible, but if the supply is from conventional sources, then a small amount of CO_2 is emitted.

Agriculture

Greenhouse cultivation: Typically, more than half of geothermal energy consumption is for space heating, while another third is used for heating pools. The remainder is used to support industrial and agricultural applications, such as greenhouse cultivation. Greenhouse cultivation has become a global industry for growing vegetables, fruits, and flowers. The process involves regulating temperature and humidity in the greenhouse to provide a conducive environment for crops. The result is a higher crop production rate, better quality, and reduced incidence of disease.

Plants need a specific temperature and humidity level for growth. For example, plants that grow in tropical climates may not be able to sustain growth in cold climate regions and vice versa. The advantage of a greenhouse is that the plant growth is not restricted to any season, hence the required food can be grown at any time of the year for domestic consumption as well as for commercial export. The Chena greenhouse in Chena, Alaska is an example of growing crops in unlikely places. The facility has been producing hydroponic lettuce, herbs, tomatoes, and small fruits since 2004. During winter, radiant air heat exchangers are used to heat the natural cold air using geothermal energy. The warm air is then ventilated into an interior greenhouse area.

Another example of greenhouse cultivation using geothermal energy includes Gourmet Mokai, New Zealand, a glasshouse facility for growing tomatoes and capsicum. The use of geothermal enables crops to be grown during the winter months, ensuring year-round production. Arataki Honey, a honey producer located in Rotorua, also in New Zealand, has used geothermal energy for honey processing for several years, resulting in cost-effective operations. The company produces more than 1,000 tonnes of honey annually for domestic and export markets.

The greenhouse heating system is crop-specific. The crop grown, the common diseases that attack the crops, the humidity requirement of the crop, and circulation air to control leaf mildew are some of the factors that guide the construction of greenhouses. Crops and plants that grow in the tropics and subtropics may require high humidity and high soil temperatures. Certain flowering plants may need shading to control blooming. The rose cultivation in Kenya is an excellent example of geothermal greenhouse cultivation. Kenya has large geothermal-based rose cultivation for commercial purposes in Olkaria. One bore well drilled to a depth of 1.6km producing 51 T/hr of hot water and steam is completely dedicated to rose cultivation and the flowers are exported worldwide. Among the countries that use geothermal resources to heat greenhouses are China, Hungary, Iceland, Italy, Kenya, the Netherlands, the Russian Federation, Türkiye, and the United States.

One method of carrying out greenhouse heating is by circulating air over finned coil heat exchangers carrying hot water in plastic tubes running along the length of the greenhouse. This maintains uniform heat throughout the length of the greenhouse. A second method is laying out pipes carrying hot water over the floor of the greenhouse. Additionally, finned units located along the walls or below the benches can be used. Of course, a combination of all the above-mentioned methods can be used.

The most economical and efficient greenhouses are large structures covering a large area, say about 36x110m constructed with fiberglass with furrow-connected gables. Heating would be through a combination of fan coils connected in series with a network of horizontal pipes installed outside walls and under the benches. A storage tank to store geothermal water is necessary to meet any peak demand about 6.3L of 60-80°c water will be necessary for peak heating. A typical geothermal-supported greenhouse is shown below:

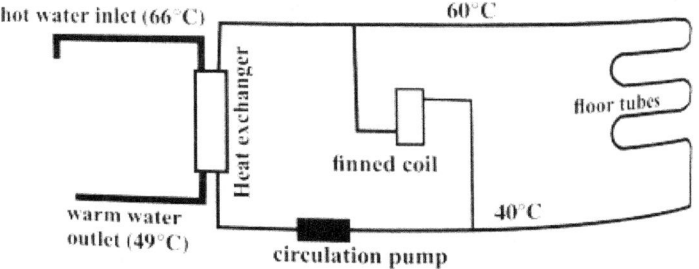

Geothermal water circulation system. source: lund, 1996

Geothermal energy can support agricultural production in various ways.
In post-harvest preservation of produce, geothermal energy can be used to support drying, dehydration, and milk pasteurization.

Drying and Dehydration are among the two most prominent uses of geothermal energy in the agri-food sector. One way for farmers to reduce waste and ensure food availability throughout the year is through the drying of agricultural products (fruits, vegetables, fish, meat, cereals, etc.). The heat required for the drying process can typically be obtained from hot water from geothermal sources. A variety of crops are suited for drying and dehydration using temperatures that can be supplied by geothermal energy. These include cereals (maize, rice, wheat, etc.), tomatoes, onions, garlic, carrots, mushrooms, apples, mangoes, pears, and dates, among others. One of the more experienced

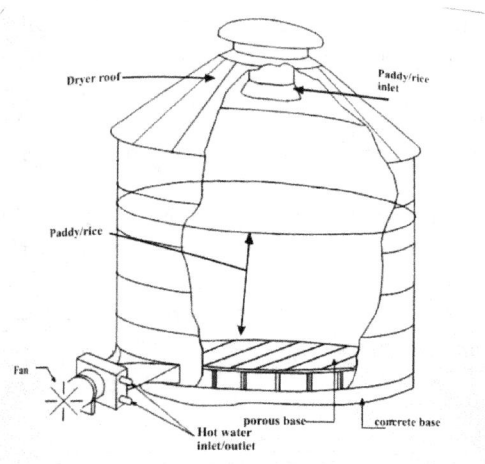

Grain dryer for removing the moisture from the food grains using geothermal heat. Source: Lund, 1996

food dehydrator plants is at Los Azufres, Mexico in operation since 1995. The facility has the capacity to dehydrate an average of 400 kilograms of fruit and produce 40 kilograms of dry fruit. The temperature inside the dehydrator averages about 60°C.

Milk pasteurization: Milk is a product that can spoil quickly due to its enzymatic activity and microbial growth and thus requires timely processing to keep it fresh. The dairy industry can use geothermal hot water for milk pasteurization as well as geothermal steam for ultra-high temperature (UHT) pasteurization and milk powder production through evaporation and drying. Geothermal heat can be used at around 60-80°C to pasteurize milk and get rid of most microbes. Most geothermal resources can provide this kind of temperature. The issue that requires more attention is cooling the treated milk to around 3-4°C for storage. This is achieved by using cold water.

Milk processing using geothermal heat has been used in Klamath Falls, Oregon USA and in Iceland since the early 1900s. Geothermal water with a temperature of 87°C and flow rate of 119L/s has been utilized for the pasteurization of milk in Klamath Falls. The geothermal water at 87 degrees is passed through a plate exchanger where the milk is heated from 3 to 71 degrees. The milk is then passed through the homogenizer and then passed through a second section of the heat exchanger where the milk is heated for 15 seconds at 77 degrees. Then the milk is culled to 12 degrees and finally chilled to 3 degrees by cold water. The advantage of this process is that the milk retains its flavor, and its shelf life is enhanced. The milk is processed at the rate of 0.8 L/s.

A more modern facility is the Maori milk processing plant in Miraka, New Zealand. The plant runs on geothermal electricity and uses geothermal heat from the Mokai Geothermal Field in its process heating applications. The plant produces dried milk powders and ultra-high temperature (UHT) products (>135°C to preserve milk for longer). These products acquire clean, green branding and are exported to more than 26 countries. The milk processor receives milk from 110 local farms and around 60,000 cows, thereby providing the milk farmers with a market for their produce. The factory provides jobs to around 120 employees who contribute to local economic growth (Wairakei Research Centre, 2020). Similar projects exist in Italy, Honduras, and Kenya

Milk with a longer shelf life can also be produced by pasteurizing the milk using higher-temperature geothermal fluids (>130°C). In this case, the milk can be stored for several weeks after it is brought to the market. On the other hand, milk powder production requires high-temperature geothermal steam (>200°C).

Spas and balneology: Many thermal springs are used at tourist resorts for recreation. Thermal springs are also used at health resorts for curing skin and bone ailments. Low-temperature waters are either directly used or mixed with cold water to suit body temperature. In addition to natural springs, the tail water of binary cycle power plants can be utilized for bathing and swimming before the water is injected into the injection well. There are several examples and sites across the world where such facilities exist, including in Iceland. The table on the left illustrates the direct utilization of geothermal energy in the world over 10 years. Heat pumps, such GSHPs, are overwhelming where direct geothermal energy is used. Followed by consumption for heating/cooling and greenhouse cultivation.

Direct use of geothermal energy (MWt) during a ten year period

Category (MWt)	2015	2005
Heat Pumps	49,898	15,384
Space heating & cooling	7,916	4,537
Greenhouse cultivation	1,830	1,404
Agriculture (drying)	161	157
Aquaculture	695	616
Industrial use	610	484
Batlung/Balneology	9,140	5,401

source: Lund and Boyd, 2015

Primary production	Post-harvest and storage	Transport and distribution	Processing	Retail preparation and cooking
• Water for irrigation • Heating of greenhouses and soil warming • Aquaculture heating • Sterilisation of soil, irrigation water and substrate for mushroom culture • Enhancing photosynthesis through CO_2 from geothermal sources • Fertiliser manufacture from sulphur • Running of water pumps using geothermal electricity	• Drying and dehydration of grains, fruits, vegetables, meat and fish, etc. • Cold storage and refrigeration (electric and thermal driven)	• Ice generated using geothermal energy • Electric vehicles charged using geothermal energy	• Process heating applications • Pasteurisation, e.g. milk • Sterilisation, e.g. food canning • Fermentation and distillation, e.g. beer, wines and spirits • Evaporation, e.g. milk powder • Powering of processing equipment using geothermal electricity	• Pre-cooking, e.g. food canning • Baking

Geothermal applications in agri-food value chains

Adapted from IRENA, 2019.

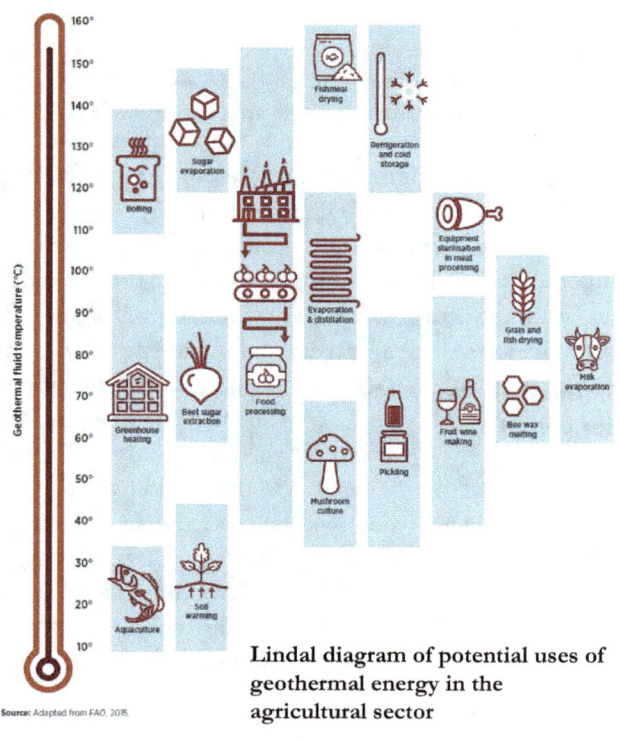

Lindal diagram of potential uses of geothermal energy in the agricultural sector

Source: Adapted from FAO, 2015.

Renewable and sustainable

Geothermal energy is considered renewable because heat is continuously being produced below the earth's surface. The energy output can be sustainable if an equal volume of water is injected back into the area where the heat extraction takes place. Thus, if high-temperature water/steam is extracted from the underground reservoir, cooler water must be replenished back to the underground reservoir. The thermal/heat difference between the waters is the energy gained.

The extraction must still be monitored to avoid local depletion. Depletions occur when steam and water are extracted faster than they are replenished. The three oldest sites, Larderello, Wairakei, and the Geysers, have experienced reduced output because of local depletion. For example, the Geysers geothermal complex in California was producing 1.5GW of electricity with 22 power units during their peak. Today, they are reduced to 900MW with 18 units because of depletion. Again, depletion can be corrected by injecting more water back into the reservoirs where steam and water are being extracted from. This method of recovery has been successfully implemented at some sites, including the Lardarello field in Italy, the Wairakei field in New Zealand and the Geysers in California. All three have been operational since 1913, 1958, and 1960 respectively. Through proper reservoir management, the rate of energy extraction can be balanced with a reservoir's natural heat recharge rate.

In addition to being renewable and sustainable, geothermal energy has the following benefits:

- **Baseload**: Geothermal power plants produce electricity consistently regardless of weather conditions.

- **Small Footprint**: Geothermal power plants are compact, using less land per GWh (404 m^2) than coal (3642 m^2), wind (1335 m^2) or solar PV (3237 m^2). Geothermal also has minimal land and freshwater requirements (5.3 US gal) than other plants like nuclear, coal, or oil (260 US gal) per MWh.

- **Clean**: Modern closed-loop geothermal power plants emit no greenhouse gases; life cycle GHG emissions are four times less than solar PV and six to twenty times lower than natural gas.

- **Scalability**: Geothermal power is highly scalable, operating from a rural village to an entire city. The power generation is local and requires no lengthy and expensive grid system.

Environmental Concerns: Although the low emissions of geothermal energy are considered to have excellent potential for the mitigation of global warming, there are some environmental effects to keep in mind. They include:

- Geothermal wells release greenhouse gases trapped deep within the Earth, although the emissions are much lower than fossil fuel emissions.

- Fluids drawn from deep Earth carry a mixture of gases, notably carbon dioxide (CO_2), hydrogen sulfide (H_2S), methane (CH_4), and ammonia (NH_3). These pollutants contribute to global warming, acid rain, and noxious smells if released. Plants that experience high levels of acids and volatile chemicals are usually equipped with emission-control systems to reduce the exhaust.

- Hot water from geothermal sources may also contain trace amounts of toxic elements such as mercury, arsenic, boron, and antimony. These chemicals precipitate as the water cools and can cause environmental damage if released.

- The required hardware, such as pumps and compressors, may consume energy from a polluting source but is a fraction of the heat output.

- Enhanced geothermal systems can trigger earthquakes as part of hydraulic fracturing. For example, a project in Basel, Switzerland, was suspended because more than 10,000 seismic events measuring up to 3.4 on the Richter scale occurred over the first six days of water injection.

As shown above, geothermal energy requires careful planning to be a sustainable and environmentally safe energy source. Also, since it is only profitably available in a few places on earth, it may never be a significant energy source for many. Seven countries are responsible for nearly 80% of global geothermal power plant capacity. However, geothermal energy can be part of a renewable energy portfolio mix. Considering geothermal energy is available under any weather conditions day and night, it can also be used as energy storage to supplement other sources such as wind and solar.

Bioenergy

Bioenergy is renewable energy derived from biological sources. It is a form of renewable energy that is derived from recently living organic materials known as biomass. Biomass is any organic material that has stored sunlight in the form of chemical energy. In simpler terms, Biomass is the fuel, and bioenergy is the energy extracted from the biomass/fuel. Biomass includes wood, wood waste, crops, crop residue like straw, food waste, animal byproducts/manure, sugarcane, and many more agricultural products and agricultural residues.

Biomass fuel was initially produced as a byproduct, residue, or waste-product of other industrial processes, such as farming, forestry, and food processing. Recently though, biomass fuel has been agriculturally grown specifically for biofuel production. These agricultural products include corn and soybeans in the United States; rapeseed, wheat, sugar beet, and willow in Europe; sugarcane in Brazil; palm oil in Southeast Asia; sorghum and cassava in China; and jatropha in India. Additionally, other biomass fuel sources are being developed. These include municipal and household waste, also known as Sewage biomass.

Biomass is mainly used in three different forms:

> **Biopower:** fuel for power plants to generate electricity
> **Biofuel**: fuel for the transportation industry
> **Bioproducts**: starter ingredient for the manufacturing industry

Biopower for power utilities

Biopower technologies convert renewable biomass fuel into heat and electricity using steam and turbine processes, similar to fossil fuel-based power plants. The biomass used for electricity production ranges by region. For example, forest byproducts, such as wood residues, are popular in the United States, while rice husks are common in Southeast Asia. Animal husbandry residues, such as poultry litter, are widespread in the UK while Brazil prefers a much abundant source of sugar cane. There are three ways to harvest the energy stored in biomass to produce biopower: burning, bacterial decay, and conversion to a gas or liquid fuel.

Burning: Most biopower is produced by burning biomass to produce high-pressure steam. Steam is used to turn a turbine attached to an electricity-generating generator in a similar fashion to fossil-fueled power plants. For example, sugar and ethanol-producing plants burn the sugarcane byproduct/waste, bagasse, to provide heat for distillation. This allows the plants to be energy self-sufficient and offset the need for carbon fuel lowering the carbon intensity of electricity generation along the way. Excess bagasse not used as fuel is used to generate electricity, which provides additional income to the plants and additional power for utilities.

Bacterial Decay: This technology uses waste material, such as human sewage or animal dung decomposed by bacteria to produce methane gas. The gas is used to replace natural gas in generating electricity.

Conversion to gas or liquid fuel: This system exposes solid biomass to high temperatures to produce synthesis gas which is a mixture of carbon monoxide (CO) and hydrogen (H). The gas is then used in a boiler to produce electricity. Another method is to heat the solid biomass to lower temperatures under a complete absence of oxygen to produce crude-like fuel. The crude-like fuel is then used to substitute for other fuel in the turbines.

Biofuel for the transportation industry

Liquified natural gas for powering commercial vehicles

As mentioned earlier, the purpose of this book is how to utilize renewable energy for electricity production. Thus, the discussion of renewable energy for the transportation industry is minimal. However, brief discussions of bioenergy for transportation are necessary, considering biofuel's success in the last few decades.

Biomass that is converted into liquid fuel or gaseous fuel is known as biofuel and is used for transportation. The two most common types of biofuel today are ethanol (Bioethanol) and diesel (biodiesel). They are mainly used in airplanes, light and heavy vehicles, and ships. Biofuel offers renewable transportation fuels that are functionally equivalent to petroleum fuel and help lower carbon intensity. Biofuel can be produced from plants, crops, commercial waste, and domestic waste. If the biomass used in the production of biofuel can regrow quickly, the fuel is generally considered to be a form of renewable energy. However, whether it is environmentally friendly or not depends on many factors, which will be discussed later in the chapter.

Bioethanol: is the most common biofuel worldwide, particularly in the United States and Brazil. It is an alcohol made by fermentation, mostly from carbohydrates produced in sugar or starch crops such as corn, sugarcane, or sweet sorghum. The process involves enzyme digestion to release sugars from stored starches, fermentation of the sugars, distillation, and drying. The distillation process requires significant energy input for heat. The process is considered renewable only if the energy input is derived from easily replenishable sources, such as trees, grasses, bagasse, and wood chips, rather than fossil fuel.

Ethanol can be used as a fuel for vehicles in its pure form (E100). However, since ethanol has a smaller energy density than gasoline, it takes much more ethanol to produce the same amount of work as gasoline. Thus, ethanol is usually used as a gasoline additive to increase octane and improve vehicle emissions. After all, ethanol does have a higher octane rating than gasoline. In 2019, worldwide biofuel production reached 161 billion liters (43 billion gallons US), contributing to 3% of the world's fuel for road transport. For those who dare to venture ahead and use biofuel for their vehicles, there is what is known as Drop-in biofuel. Drop-in biofuels are functionally equivalent to petroleum fuel and fully compatible with the existing petroleum infrastructure. They require no engine modification of the vehicle.

Biodiesel: is produced from oils or fats using transesterification and is the most common biofuel in Europe. Chemically, it consists mostly of fatty acid methyl (or ethyl) esters (FAMEs). Feedstocks for biodiesel include animal fats, vegetable oils, soy, rapeseed, sunflower, palm oil, and algae, to name a few. It can be used as a fuel for vehicles in its pure form (B100) but is usually used as a diesel additive to reduce particulates, carbon monoxide, and hydrocarbons from diesel-powered vehicles to improve efficiency. In many European countries, a 5% biodiesel blend is widely used and is available at thousands of gas stations. In the US, more than 80% of commercial trucks and city buses run on diesel. Biodiesel is also safe to handle and transport because it is non-toxic and biodegradable.

Green diesel: Green diesel is produced through hydrocracking biological oil feedstocks, such as vegetable oils and animal fats. Hydrocracking is a refinery method that uses elevated temperatures and pressure in the presence of a catalyst to break down larger molecules, such as those found in vegetable oils, into shorter hydrocarbon chains used in diesel engines. Unlike biodiesel, green diesel has precisely the same chemical properties as petroleum-based diesel. It does not require new engines, pipelines, or infrastructure to distribute and use. Unfortunately, it has not been produced at a cost that is competitive with petroleum. Green diesel is being developed in Louisiana and Singapore by ConocoPhillips, and Valero among other companies.

Straight vegetable oil: Straight unmodified edible vegetable oil is generally not used as fuel, but this truck is one of 15 based at Walmart's Buckeye, Arizona distribution center that was converted to run on a biofuel made from reclaimed cooking grease produced during food preparation at Walmart stores.

Bioproducts for the manufacturing industry

In addition to being used for biopower and biofuel, biomass can also serve as a renew- able alternative to fossil fuel in manufacturing bioproducts such as plastics, lubricants, industrial chemicals, and many other products currently derived from petroleum.

Investment and jobs

The economic contribution of bioenergy, particularly in the generation of electricity, is still minimal. However, bioenergy can be lucrative in an isolated situation where waste byproducts are recycled for fuel. Deforestation and agricultural biomass competing for arable land, however, have unforeseen, long-term consequences for the overall environment. U.S. Department of Energy's 2016 Billion-Ton Report: Advancing Domestic Resources for a Thriving Bioeconomy concluded that the United States has the potential to produce 1 billion dry tons of non-food biomass resources annually by 2040 and still meet demands for food, feed, and fiber. One billion tons of biomass could:

- Produce up to 50 billion gallons of biofuel
- Yield 50 billion pounds of bio-based chemicals and bioproducts

- Generate 85 billion kilowatt-hours of electricity to power 7 million households
- Contribute 1.1 million jobs to the U.S. economy
- Keep $260 billion in the United States.
- By 2010, there was 35 GW (47,000,000 hp) of globally installed bioenergy capacity for electricity generation, of which 7 GW (9,400,000 hp) was in the United States.

Bioenergy and the environment

The burning of carbon-based fuel always leads to carbon-based emissions regardless of whether the carbon originated from fossil fuel or biofuel. Unfortunately, many biofuel projects are not carbon neutral. Some even have higher carbon emissions than fossil-based projects. In general, any fuel or energy is considered a pollutant when released into the environment at a rate faster than the environment can disperse, dilute, decompose, recycle, or store it in some harmless form. Based on this definition, both fossil fuels and some biofuels are pollutants. Thus, choosing the proper biomass is critical to adding bioenergy to our renewable energy portfolio.

For instance, in 2018, the European parliament voted to phase out palm oil use in transport fuel by 2030. A 2015 study funded by the European Commission found that palm oil and soybean oil had the highest indirect greenhouse gas emissions due to deforestation and drainage of peatlands.

Increased logging and deforestation are also taking place globally to support the bioenergy industry. The long-term consequences of deforestation are scientifically acknowledged. Forests are responsible for consuming 25% of the CO_2 humans produce.

Agricultural biomass, competing with food production for arable land, is an increasing concern as well. To calculate land use requirements for different kinds of power production, it is essential to know the relevant area-specific power densities. Smil estimates that the average area-specific power densities for biofuel, wind, hydro, and solar power production are 0.30 W/m^2, 1 W/m^2, 3 W/m^2, and 5 W/m^2, respectively (power in the form of heat for biofuel, and electricity for wind, hydro and solar). The average human power consumption on ice-free land is 0.125 W/m^2 (heat and electricity combined), although rising to 20 W/m^2 in urban and industrial areas. The low yields of biofuel make it unattractive compared to other renewables.

Conclusion

The climate change and energy debate has produced two passionately vocal sides. The opponents of fossil fuel-based economies are convinced of the negative effects of fossil fuels on our environment and our health and are promoting change in the energy industry. The proponents of the fossil fuels industry, on the other hand, are not convinced by the climate change argument and thus, are protective of the status quo. They argue that climate change research, data, and analysis are misunderstood and misrepresented. They insist that the issue is being politicized against communities and nations. The renewable energy industry can bring both camps under one tent. It offers a cleaner and healthier option while providing cheaper and abundant sources of energy to support growing population and economies.

Fossil fuels have been an integral part of the global economic engine for the last two centuries. All industrial revolutions beginning with the first in the late 1700s, were made possible by fossil fuels. Coal, oil, and natural gas have all contributed to the development of the manufacturing, transportation, and utility industries. Currently, fossil fuels account for 60% of the U.S. and 80% of the global energy consumption. Furthermore, the use of fossil fuels has been increasing at record levels in recent decades.

Understandably, minimizing or abandoning the use of fossil fuels is not an easy task. How does one convince nations to forgo the time and financial commitments they had invested in their fossil fuel-based economies and promote renewable energy sources that still seem underdeveloped? How does one convince nations like Nigeria and Saudi Arabia, who are blessed with fossil fuel resources, to forgo their advantages and budget for wind turbines and solar panels from abroad? How does one convince China which relies on fossil fuels for over 60% of its economy to shut down its coal power plants and allocate new budget for other sources? It is obvious that governments have the responsibility of improving the livelihood of their population and will use whatever resources are at their disposal to achieve their goals. Thus, the climate change argument alone cannot convince the world to change energy consumption behavior. As a matter of fact, any rush judgment or fear-mongering against fossil fuels can be economically costly.

However, research and analysis, shared over the past few decades, stress the negative impact of the continued use of fossil fuels. Worsening environmental conditions and degrading public health are the main concerns. The rising cost and dwindling availability of fossil fuels are additional concerns. The use of fossil fuels as a weapon in destabilizing international economic trade is also an increasing worry. Therefore, opponents of the fossil fuel industry continue to campaign for an energy shift towards other locally available sources of energy, mainly renewable energy.

Regardless, both sides of the argument must be carefully considered. What if the climate change assessment is correct in projecting worsening climate and degraded public health? Can we afford to wait for additional research or a confirmation? On the other hand, does championing renewable energy lead to the opportunity cost of continuing to use fossil fuels? what if we accept the climate change research and analysis and start to gradually shift towards renewable energy? Can the renewable energy industry provide competitive energy pricing and create employment opportunities?

As we start to fully understand what we are asked to change, there will be less controversy and more excitement for change because renewable energy offers cleaner and more sustainable options. First of all, these resources have been used for thousands of years. With increasing technology, they are being developed further. Over the past decades, technological advancements, government policies, and consumer awareness have played a significant role in advancing the renewable energy cause in the United States and worldwide. For example, the Solar Investment Tax Credit (ITC) passed by the U.S. Congress in 2006 has significantly contributed to solar energy growth in the United States. Additional government programs, such as the U.S. "Energy Star" program, have focused on educating the public on the benefits of energy-efficient products. The energy efficiency sector has saved and continues to save Americans $billions in utility bills.

The tax credits for wind and solar energy have also contributed to a remarkable drop in the cost of renewable projects. In the past decade, wind energy costs have declined nearly 70 percent, while utility-scale solar prices fell by almost 90 percent. Solar and wind power are now cheaper sources of electricity than coal, nuclear, and natural gas technologies. For the first time, renewable energy sources can be evaluated on an equal footing with traditional fossil fuel-based energy. This is extremely important considering the percentage of income spent on utility bills. For example, in the United States, over a third of household expenses are spent on heating, cooling, and powering homes. For commercial and industrial facilities, the cost is even higher. At a time when fossil fuel costs are increasing and fossil fuel availability is dwindling, the cost of energy is only going to increase.

The renewable energy industry has also helped usher in billions of dollars in investments and job creation for local economies. On a localized level, many states, counties, municipalities, and utilities are implementing policies and incentives to develop their renewable energy industries further. As a result, the industry is creating more employment opportunities than the fossil fuels industry. More importantly, the economic gains are benefiting local communities.

Furthermore, renewable energy can help curb global conflicts by allowing nations to become self-sufficient. Since the creation of the United Nations in 1945, there have been many interstate and intrastate treaties to minimize global conflicts for energy access. In addition to the efforts of the UN, developing the domestic energy sector can play a crucial role and contribute to a peaceful resolution of some global conflicts. Prioritizing energy supplies based on domestically available renewable sources reduces dependency on foreign nations and contributes to energy stability. Local governments and communities also have the freedom to implement policies and actions independently.

Hopefully, we have demonstrated that renewable energy is not a new concept. It has long been used for centuries all over the world. In fact, today's modern renewable energy industry is set on a foundation of past experiences. In addition, we have demonstrated that renewable energy is the cheapest source of electricity. Most importantly, renewable energy is abundant and infinite. Therefore, if we can increase awareness, technological advancement, and investment, there is no doubt that most nations and local communities will gradually shift towards renewable energy.
The benefits of renewable energy need to be communicated so that nations and local communities can choose to champion it. The communication tool cannot be a military or political force, monetary sanctions, or similar manipulative techniques. The benefits of shifting to renewable energy should be understood and accepted voluntarily.

To those who believe the climate is changing for the worse, there are cleaner and renewable energy sources to promote. Renewable energy helps create a cleaner environment and improves public health indiscriminately. To those who deny climate change, if we are wrong about climate change, the worst that can happen is a cleaner environment and the development of additional energy sources for future energy consumption. Therefore, a gradual shift from fossil-based energy to renewable energy is economically, socially, and politically beneficial.

Renewable energy is a win-win for all

References

A new translation by Robin Waterfield, "Herodotus, The Histories" Oxford world's classic, 1998
American Clean Power Association (www.cleanpower.org) American Oil & Gas Historical Society (www.aoghs.org)
Beloved Community Initiative (www.becomingbelovedcommunity.org) Britannica (www.britannica.com)
British Library (www.bl.uk/georgian-britain)
Bureau of Ocean Energy Management (BOEM) (https://www.boem.gov) Centers for Disease Control and Prevention (www.cdc.gov)
Dakota Access Pipeline (www.daplpipelinefacts.com)
DRC (https://peacekeeping.un.org/en/mission/monusco) Earth Watch (www.earth.esa.int)
Earth works (www.earthworks.org) Eco home (www.ecohome.net)
Energy sage (https://www.energysage.com/solar/101/types-solar-panels/) Energy Start (www.energystar.gov)
Environmental and Energy Study Institute (EESI) (www.eesi.org) Eritrean center for strategic studies (http://www.ecss-online.com/) Explore the World of Piping (www.wermac.org)
Food and Agriculture Organization of the United Nations (www.fao.org)
Geo world map (https://energyeducation.ca/encyclopedia/Geothermal_district_heat-ing)
History of Wind Energy in Encyclopedia of Energy Vol. 6, page 426
https://energyeducation.ca/encyclopedia/Pump_jack
https://oilprice.com/Energy/
Global Policy forum (https://www.globalpolicy.org/security-council/dark-side-of-natural-resources/timber-in-conflict.html)
Greenpeace (https://www.greenpeace.org/usa/arctic/issues/oil-drilling/)
International Energy Agency (www.iea.org)
International Solar Energy Sociel (ISES) (www.ises.org)
King Leopold's Ghost, by Adam Hochschild, copyright 1998 by Adam Hochschild; prologue, 18
National Energy Education Development Project (NEED) (www.need.org)
National Oceanic and Atmospheric Administration (www.noaa.gov) National Park Service-U.S. Department of the Interior (www.nps.gov) National Parks Conservation Association (www.npca.org)
National Renewable Energy Laboratory (NREL) (www.nrel.gov) Natural Resources Defense Council (www.nrdc.org)

New Living Translation (www.bible.com)
New World Encyclopedia (www.newworldencyclopedia.org) Radio Free Europe Radio Liberty (www.rferl.org)
Solar Energy Industries Association (www.seia.org)
The Travels of Marco Polo (https://en.wikisource.org/wiki/The_Travels_of_Marco_Polo/Book_2/Chapter_30)
U.S. Army Corps of Engineers (www.usace.army.mil)
U.S. Bureau of Labor Statistics (BLS) (www.bls.gov)
U.S. Bureau of Reclamation (https://www.usbr.gov)
U.S. Department of Energy (www.energy.gov)
U.S. Department of Labor (www.dol.gov)
U.S. Department of Transportation (www.phmsa.dot.gov)
U.S. Energy and Employment Report (USEER) (www.usenergyjobs.org)
U.S. Energy Information Administration (www.eia.gov)
U.S. Fish & Wildlife Service (www.fws.gov)
UN Environmental Programme (www.unep.org) United Nations and the Rule of Law (www.un.org)
United States Environmental Protection Agency (www.epa.gov) United States Geological Survey (www.usgs.gov)
Wikipedia (https://en.wikipedia.org/wiki/Hydraulic_fracturing) Wikipedia (https://en.wikipedia.org/wiki/Thermal_mass) World Health Organization (www.who.int)

Photo credits

The majority of the photos and illustrations in this book are properties of the author. The photos listed below are courtesy of various individuals and agencies.

Power plant: Pixabay power-plant-1892407_1920 (pg. 16)
Strip Mining: Image by Free-Photos from Pixabay: pit 984037_1920 (pg. 16)
Pump jack: Image by mhouge from Pixabay: pumpjacks 3804889_1920 (pg. 17)
Water pollution: jwvein from Pixabay: monolithic part of the waters 3137978_1920 (pg. 20)
Theodore Roosevelt National Park, North Dakota, USA: Photo by Intricate (pg. 21)
Vertical Blades: Nashtifan, Iran: https://www.irandestination.com/asbads-of-iran/ (pg. 35)
Ireland home: Image by David Mark from Pixabay_ireland-69817_1920-1 Energy Star (pg. 58)
Energy Star pics: https://www.energystar.gov/ (pg. 63)
Pelton turbine: Wolfram Linden from Pixabay_metal-3707932_1920 (pg. 76)
Turbine cross-section: Bureau of Reclamation usbr.gov (pg. 87)
Eritrean dams: Eritrean Center for Strategic Studies (pg. 90)
Steel tower: (Joenomias) Menno de Jong from Pixabaywind-turbine-3840422_1920_mod (pg. 97)
Lattice tower: Lolame from Pixabay_wind-power-5457775_1920 (pg. 98)
Blade transport: Bishnu Sarangi from Pixabay_long-vehicle-320296_1920 (pg. 98)
Inside of wind turbine/Nacelle: U.S. Department of Energy (pg. 103)
Installation left, center, right: Erich Westendarp from Pixabay_wind-power-3045169_1920 : Hans Linde from Pixabay_windrader-2759645_1920 : Hans Linde from Pixabay_windrader-2759650_1920 (pg. 104)
Onshore wind farm: David Mark from Pixab_wind-farm-1771238_1920 (pg. 106)
HAWT: (Joenomias) Menno de Jong from Pixabay_eco-friend- ly-2232415_1920 (pg. 110)
VAWT: Rafael Albaladejo from Pixabay_energy-4734634_1920 (pg. 111) Solar panel details: Based on pic from www.solarchoice.net.au (pg. 117)
Panel: Sebastian Ganso from Pixabay_photovoltaic-system-2742302_1920 (pg. 117)
Periodic Tbl: ExplorersInternational from Pixabay_science-2227606_1920 (pg. 121)
Fresnel-Collector: U.S. Department of Energy: Ferrostaal-Hauke-Dressler (pg. 126)
Parabolic: U.S. Department of Energy (pg. 126)
Danakil region: Afrikit from Pixabay_ethiopia-634222_1920 (pg. 134)

www.ingramcontent.com/pod-product-compliance
Lightning Source LLC
Chambersburg PA
CBHW052255220526
45471CB00001B/346